THAT'S THE WAY
THE COOKIE CRUMBLES

THAT'S THE WAY THE COOKIE CRUMBLES

62 All-New Commentaries on the Fascinating Chemistry of Everyday Life

DR. JOE SCHWARCZ

Director
McGill University Office for Science and Society

ECW PRESS

Published by ECW PRESS
2120 Queen Street East, Suite 200, Toronto, Ontario, Canada M4E 1E2

NATIONAL LIBRARY OF CANADA CATALOGUING IN PUBLICATION DATA

Schwarcz, Joseph A.
That's the way the cookie crumbles: 62 all-new commentaries
on the fascinating chemistry of everyday life / Joe Schwarcz.
ISBN 1-55022-520-0
QD37.S383 2002 540

Cover design: Guylaine Regimbald – SOLO DESIGN.
Copyeditor: Mary Williams.
Production: Emma McKay.
Interior design: Yolande Martel.
Interior cartoons: Brian Gable.
Author photo: Tony Laurinaitis.
Printing: Transcontinental.

This book is set in Stempel Garamond and Koch Antiqua.

The publication of *That's the Way the Cookie Crumbles* has been generously supported by the Canada Council, the Ontario Arts Council, and the Government of Canada through the Book Publishing Industry Development Program. Canadä

DISTRIBUTION

CANADA: Jaguar Book Group, 100 Armstrong Avenue,
Georgetown, Ontario L7G 5S4

UNITED STATES: Independent Publishers Group, 814 North Franklin Street,
Chicago, Illinois 60610

EUROPE: Turnaround Publisher Services, Unit 3, Olympia Trading Estate,
Coburg Road, Wood Green, London N22 6T2

AUSTRALIA AND NEW ZEALAND: Wakefield Press, 1 The Parade West (Box 2066),
Kent Town, South Australia 5071

PRINTED AND BOUND IN CANADA

ECW PRESS
ecwpress.com

CONTENTS

EVERYDAY SCIENCE

INTRODUCTION

Funny thing, memory. Save for some vague recollections of pulleys, pumps, inclined planes, and a cute redheaded girl, my grade-nine science class is mostly a mental blur. But one little episode that occurred way back then did manage to etch itself indelibly in my mind. I remember asking Mr. Labcoat a question, the nature of which I have long forgotten, and being surprised by his rather curious answer. "That's just the way the cookie crumbles," he blurted out. I didn't quite understand the reference to culinary chemistry, but I did understand that he had no ready answer and was unwilling to search for one.

These days, when I spend much of my day trying to answer people's questions about science, that phrase uttered so many years ago often leaps to mind. There are two basic reasons for this. First, it is a great motivator to do the necessary research instead of offering up the easy, "cookie" answer. But it is also a constant reminder of the limitations of our scientific knowledge. Why do some people experience side effects from a medication, while others do not? Why does cancer often seem to strike people who do everything right in terms of lifestyle and spare those who have a more cavalier attitude towards their health? Why do some people say that they sleep better with their feet pointed towards the magnetic North Pole? Why do they believe in such nonsense? We just don't know. Sometimes, I guess, it's just the way the cookie crumbles.

I've now been dealing with crumbling cookies, in a manner of speaking, for over twenty years. About two decades back, a couple of colleagues and I were approached to mount some sort of science display at the annual Man and His World exhibition, a descendant of Expo '67, the hugely successful Montreal world's fair. One of the demonstrations we featured was making polyurethane foam. This involved mixing two reagents in a cup, and it resulted in a mountain of foam — which, to the delight of young and old, quickly hardened into a mushroom-shaped blob. A really neat demo. We had a lot of fun with it until the proverbial fly fell into the ointment one Monday morning. I remember it well.

I picked up my newspaper and began to glance through it in the usual fashion. To my great surprise, the page 3 city column was all about our chemical escapades. It described how, in spite of the great public anxiety about urea-formaldehyde foam insulation, "some chemists" were brewing it up in front of a live audience and singing its praises. Well, that got me more than a little hot under the collar. True, there was concern at the time about urea-formaldehyde insulation, specifically about its potentially toxic formaldehyde content. But we were not dealing with urea-formaldehyde. We were demonstrating the properties of polyurethane, a distinctly different material. The only common feature was that these were both foams.

By nine o'clock that morning, I had delivered a letter to the newspaper, along with a large egg formulated out of polyurethane, which I suggested the columnist hang around his neck for penance. After all, he had laid a large egg by not appreciating the difference between urea-formaldehyde and polyurethane. His misunderstanding, I went on to say, had undoubtedly created unnecessary concerns. Much to his credit, the columnist wrote a retraction, explaining that the real problem was his lack of scientific education, which had caused him to leap to inappropriate conclusions. I was satisfied and thought the issue closed. But then I got a call from a local radio station asking if I would like to comment on this controversy, which of course was really a noncontroversy. The people at the station must have liked the way I explained the matter, because a couple of weeks later they again asked me to go on the air and discuss some chemistry-related issue that had arisen. Soon, this evolved into a series of spots, eventually leading to a regular weekly phone-in show entitled *The Right Chemistry*, which continues to this day.

The radio program spawned requests to give public lectures and make television appearances, as well as invitations to write newspaper columns and books. In 1999, these efforts culminated in the establishment of the McGill University Office for Chemistry and Society. The goal of this unique venture is to increase the general public's understanding of, and appreciation for, science. Now expanded and renamed, the Office for Science and Society aims to provide accurate, unbiased scientific information on various issues of public concern and welcomes all kinds of queries about scientific matters, particularly as they pertain to daily life.

Interacting in this fashion with the public over so many years has been a fascinating, exciting, fulfilling, and sometimes frustrating experience. Above all, it has afforded insight into the public's fears, concerns, hopes, and dreams, both rational and irrational. Anyone involved in this business quickly realizes that there are numerous misconceptions about science out there that need to be addressed. It has also become painfully clear that whenever science cannot provide an adequate answer, charlatans rush in to fill the void. This volume — like its predecessors, *Radar, Hula Hoops, and Playful Pigs*, and *The Genie in the Bottle* — aims to educate and entertain the reader with up-to-date, readily understandable commentaries designed not only to help develop a feel for the workings of science, but also to provide some of the background needed to separate sense from nonsense. And there's plenty of down-to-earth, practical scientific information here as well. You'll learn how to remove stains from clothes, how to lower your cholesterol with oats, how to make "oobleck" — and you'll discover why the cookie crumbles.

HEALTHY SCIENCE

Microwaved Socks, and Other Tales from the Airwaves

I look forward to my Friday mornings. That's when I spend an hour and a half conversing with the public on CJAD Radio in Montreal. The idea behind the show is for me to provide reliable scientific information, answer questions about current concerns, and attempt to clear up some of the mysteries that permeate daily life. But the show is an education for me, as well. For over twenty years, it has allowed me to monitor the pulse of the population and gain a glimpse into its psyche. I have been pleased by callers who have made unusual scientific observations, elated by those with intriguing questions, and frustrated by the occasional demonstration of scientific illiteracy. I have also come to realize that people are burdened with numerous fears, both rational and irrational. And I have learned not to be surprised. Shocked, maybe — but never surprised.

"How do you wash microwaves out of socks?" one caller queried. I didn't quite know what to make of this. Quickly, though, we established that he was not worried about having trodden on some stray microwaves, but he had heard about a device being marketed to reduce the risks of cell phone use.

First of all, we need to understand that there is very little scientific evidence to suggest that cell phones are dangerous, other than to those who use them while driving. But that has not stopped the inventive marketers. They've come up with a socklike device that one places over the phone to absorb the "harmful microwaves." The instructions that come along with this gem apparently instruct the user to launder the sock regularly to "wash out the radiation." Total nonsense.

Microwaves are a form of energy, and they can indeed be absorbed by materials. After all, that's how microwave ovens work. Moisture absorbs the waves, energizing the water molecules. They move around more rapidly, and it is this motion that we sense as heat. But microwaves cannot be stored in a substance for later release. It seems, though, that this contention is not restricted to scam artists who want to protect us from cell phones. A listener once called to ask how long one should allow microwaved food to stand after cooking to "allow the microwaves to escape." Obviously, this person had been reading her microwave cookbook, which would have advised her to allow microwaved food to stand briefly before serving it. This is common practice, necessary to complete the cooking process. Contrary to what many think, microwaves do not penetrate food deeply. The exterior of the food in question is easily heated, but the inside cooks through heat transfer by conduction. That's why the food must stand for a few minutes. It has nothing to do with allowing vagrant microwaves to escape.

Microwaves are not the only form of radiation causing undue concern. A terribly agitated caller was worried because after being x-rayed she was asked to take the films to her physician herself. She had heard all about exposure to x-rays being dangerous and thought that by carrying the films she was "exposing" herself. Since the infamous date of 9/11, a number of people

have asked about wearing clothes that have gone through x-ray scanners at airports. They are concerned that these items may become radioactive and pose a risk to their health. Excessive exposure to x-rays can certainly be risky, but x-rayed items do not store and reemit radiation. Unfortunately, just a mention of the word *radiation* is often enough to alarm people.

A gentleman wanted to know what the safest way was to dispose of a broken compact-disc player. I didn't realize what he was getting at until he asked whether a laser was a form of radiation, which of course it is. Radiation is nothing more than the propagation of energy through space. Turn on a light and you are exposed to radiation. The caller knew that CD players use a laser, and since lasers produce radiation, there had to be some risk. The laser beam in a CD player is just a special type of light beam that poses no danger at all, and it is only emitted when the device is on. So old CD players can be safely discarded. But old laminated pictures may be a different story.

I had to address this issue when a caller asked if it was safe to burn a laminated picture in her fireplace. It turned out that she had been recently divorced and wanted no reminders of her former spouse. Burning his picture seemed appropriate, but she had heard that laminated photos were mounted on particle-board glued together with urea-formaldehyde resin. She was worried that the heat would release formaldehyde, which she had heard was toxic. Indeed, formaldehyde is a problematic substance, but the amount released in this particular combustion process would be too little to cause concern. Still, I suggested that if she was still worried, she could hang on to the picture until the next hazardous waste collection took place in her municipality. She liked that idea — she told me that "hazardous waste" was an excellent description of her former mate.

Then there was the listener who wanted to know if lighting a match was a good way to get rid of the smell of natural gas in a house. That question prompted me to launch into a lecture on a common misunderstanding about gas. Natural gas, I said, is just methane, and methane has no smell. That's why odiferous compounds are added to make sure that gas leaks are readily detected. I explained that soot from a burning match could absorb small amounts of smelly compounds, but, I added some-what smugly, it was not a good idea to go around striking matches in a house that could be filled with methane. It was then that the caller sheepishly informed me he knew all that, but the "natural gas" he was talking about was more likely to be found in the bathroom than in the kitchen. It was I, not he, who had jumped to the wrong conclusion. Like I said, my Friday mornings are interesting.

YES, SCIENTISTS ARE ALLOWED
TO CHANGE THEIR MINDS

"Gee! You scientists — one day you say this, the next day you say that," exclaimed the apparently frustrated lady as she approached me after a public lecture. "I remember having you as a prof twenty years ago, and you maintained that anyone who had a balanced diet did not have to worry about taking vitamin supplements. Now here you are suggesting that we take a multivitamin and that there may be value in some other supplements as well. Can't you guys make up your minds?"

Well, frankly, no. At least not completely. Science rarely gives us conclusive answers. It is an ongoing process that attempts to remain in step with the latest research. That's why changes in recommendations made to the public should come as no surprise. Indeed, I think I would be more concerned if I were saying the same things today as I did twenty years ago, because it would indicate that we had made no progress in our understanding of nutrition. Also, I would suggest that a time span of twenty years is not exactly "one day this, the next day that." During this period, for example, we have firmly established the role of the B vitamins in preventing the buildup of homocysteine in the blood, and we have recognized the potential value of an increased intake of vitamin D. In general, we have come to understand more fully that some vitamins may do more than just prevent nutritional deficiency diseases. At worst, as I've often said, a multivitamin may just make for more expensive urine. At best, it may result in significant health benefits.

That's the argument I was giving to my former student when another gentleman joined the discussion. He took issue with what he perceived to be my enthusiasm for vitamin supplements (although I hardly think that suggesting a multivitamin qualifies me as a supplement pusher) and informed me that he had just given up his vitamin E pills because he did not want to "die of a heart attack." I was a little taken aback by this, because vitamin E has been associated with protection against heart disease. So what was this guy talking about? He had "proof" of the dangers of vitamin E, he said, and he proceeded to pull a neatly folded newspaper article out of his pocket. "Vitamin E Debunked as Heart Healthy," the headline screamed.

Now it all began to click. A study published in *The New England Journal of Medicine* had garnered a great deal of media attention because it suggested that taking a mixture of antioxidants reduced the effectiveness of a cholesterol-lowering medication. It was an interesting study, to be sure, but it certainly did not show that vitamin E supplements are dangerous, as many newspaper accounts inferred.

One of the most commonly prescribed classes of drugs for reducing LDL cholesterol, the so-called "bad cholesterol," is the statins. LDL causes deposits to form in arteries and triggers heart attacks. This happens when the cholesterol that has been deposited undergoes a chemical reaction and becomes oxidized cholesterol, the most dangerous form. By contrast, HDL, or high density lipoprotein, scavenges cholesterol from the blood and prevents its deposition. That's why it is referred to as "good cholesterol." HDL can be increased by taking high doses of niacin, one of the B vitamins. There is also some evidence that antioxidants such as vitamin C, vitamin E, beta-carotene, and selenium can reduce the oxidation of LDL and therefore reduce the risk of heart disease. A logical question to ask, therefore, is what would happen if someone at risk for heart disease took a statin drug, plus niacin, plus antioxidants? Would the beneficial effects be combined and the risk reduced accordingly?

This is just what researchers at the National Institutes of Health decided to find out. So they enrolled 160 subjects with existing heart disease in a study. Forty patients were given

Zocor — one of the most popular of the statin drugs — together with niacin, while forty others took the same drugs plus vitamins C, E, beta-carotene, and selenium. Another group of forty took only the antioxidants, and a fourth group was given a placebo. The results were surprising. As expected, Zocor and niacin reduced LDL, raised HDL, and lowered the risk of heart attack. But when researchers incorporated the antioxidants into the mix, they noted a smaller increase in HDL and a slight increase in artery blockages. The groups taking only the antioxidants or a placebo had the most cardiac complications.

What are we to make of this? The basic conclusion is that the mix of antioxidants blunted the benefits of the combined statin-niacin therapy. We don't know which antioxidant was responsible. Vitamin E may not even have played a role at all. And this study doesn't give us information about what happens when only a statin is taken together with antioxidants, a situation that is far more common than combined statin-niacin therapy. Finally, we must remember that all the participants in this trial already had heart disease, so we can draw no conclusions at all about the possible effectiveness of antioxidants in preventing the disease in the first place. Certainly, the data do not justify headlines about antioxidants increasing the risk of heart disease.

Really, the only appropriate conclusion we can draw from this study is that anyone who has been prescribed both a statin and niacin is probably better off not taking any antioxidants. But there is absolutely nothing in the study to suggest that antioxidants, and vitamin E in particular, are dangerous to the general population. Any contention about vitamin E causing heart attacks is totally unjustified. True, the study does hint that the benefits of antioxidant supplements have probably been overstated. We'll see. No one can be certain about what further research will show. But of one thing, I am sure. If I'm around in twenty years to talk about this stuff, I won't be saying the same things as I'm saying now. That's the way science works. And do I have a better chance of being around if I take antioxidants? I guess that depends on which study I read.

I'll Take the Yogurt,
but Hold the Enema Machine

It was an experience I never thought I would have. I had read about it, I had seen pictures of it. Now here I was, sitting in Dr. John Harvey Kellogg's patented vibrating chair, happily vvvibrrraaatinggg away. I was visiting the little museum in Battle Creek, Michigan, dedicated to the exploits of the good doctor, whose flaky ideas about health had captured America's imagination in the late 1800s.

Kellogg was infatuated with the human colon. He believed that virtually all ailments could be traced to "autointoxication" through substances produced by the "putrefying" bacteria inhabiting the colon. The key to health, he maintained, is a clean colon. His cereal flakes served as "little internal brooms" that helped sweep out the colon's contents, especially if these were loosened by the activity of the vibrating chair. I didn't note any such effect. Neither did I notice any stimulation of the internal organs after galloping on Dr. Kellogg's famous mechanical horse. But I did spy something while riding that horse. A curious glass and metal device sat unceremoniously in a corner, barely eliciting a glance from other museum visitors. I recognized it right away. It was Kellogg's enema machine.

Dr. John Harvey Kellogg administered and received more enemas than anyone in history. His electric enema appliance pumped fifteen gallons of water through the colon in just one minute. But that was not the end of the cleansing process. Next came the yogurt flush. Made with the amicable bacterium *Lactobacillus bulgaricus*, the yogurt would "drive out the disease-forming bacteria that had been implanted by the putrefactive action of flesh foods." "Balance your intestinal flora," Kellogg maintained, "and you'll live as long as the rugged mountain men of Bulgaria!" And, according to Elie Metchnikoff, the Russian bacteriologist whose research triggered Kellogg's yogurt compulsion, that was pretty long.

Metchnikoff had caused quite a sensation with his theory that the longevity of Bulgarians was due to the copious amounts of yogurt they ate. He even had a theory to explain how this happened. The good bugs, which Metchnikoff named in honor of the Bulgarians, overwhelmed the bad bugs in the gut that caused disease. Yogurt was elevated to the rank of a wonder food, in spite of the fact that Metchnikoff had no real evidence for his theory, or indeed for his notion that Bulgarians experienced remarkable longevity.

Metchnikoff was awarded a Nobel Prize in 1908 (for work unrelated to yogurt), and that helped enshrine the yogurt mystique and make the food a "health" staple in the Balkans and in Russia. Indeed, when the former Soviet Union began its manned space flight program, it established a microbiology laboratory at the Baikonur Cosmodrome to study the cosmonauts' gut bacteria. The concern was that the stresses of space travel might change the balance of these bacteria and cause some nasty symptoms. A spacecraft is certainly no place to be struck by diarrhea!

The researchers experimented with giving the cosmonauts yogurt before their missions. They must have been satisfied with the results, because the practice became routine, as did the collection of bacteria samples from the spacemen's saliva and guts after they returned to Earth. The researchers cultured these samples and used them to make yogurt with the hope that, having withstood the stresses of space travel, these bacteria would create a healthier product. They probably didn't, but they did help the struggling Russian space program raise funds. A commercial variety of yogurt made with bacteria cultured from cosmonaut emissions is still being touted as a health food.

While the appeal of yogurt cultured from cosmonaut poop may be limited, the notion of introducing beneficial bacteria into the gut is receiving widespread attention from scientists. Research into "probiotics" is mushrooming. Simply put, a probiotic is any preparation that contains specific microorganisms in sufficient numbers to alter the microbial flora in a host and exert beneficial health effects. There is increasing evidence that yogurt, if made with the right bacteria, falls into this category. Traditionally, yogurt has been made with *Lactobacillus bulgaricus* and *Streptococcus thermophilus*, which are acid sensitive and do not make it through the stomach to the colon in sufficient numbers. But acidophilus and bifido bacteria do. And they really do squeeze out disease-causing bacteria, such as *Clostridium difficile*, often responsible for diarrhea.

Very good scientific evidence now exists for treating diarrhea with probiotics, including the type associated with antibiotic use. Antibiotics can destroy some of the gut's beneficial bacteria, the kind that keep troublesome microbes in check. *Clostridium difficile* can seize the opportunity, multiply, and cause severe diarrhea. In rare cases, such diarrhea resists all treatments. Well, maybe not all. Dr. Lawrence Brandt, a gastroenterologist at Montefiore Medical Center in New York, has come up with a novel, though admittedly unconventional, approach. In one highly publicized case, he mixed stool samples from a patient's husband in saline water and deposited little chunks of this matter every ten centimeters along the woman's colon. This "fecal colonoscopy" resulted in the almost immediate resolution of symptoms, in all likelihood due to the restoration of a healthy balance of microflora in the gut. Doesn't sound particularly appealing, but protracted diarrhea is no pleasure either. Science sometimes moves forward in unusual ways.

In children, diarrhea is most often caused by a rotavirus infection. Studies have shown that treatment with an oral hydration solution containing *Lactobacillus GG* significantly shortens the duration of the problem. Yet that may be just the starting point for probiotic benefits. There is tantalizing evidence for cancer prevention and immune system enhancement. Some probiotics can destroy cancer-causing agents in the gut, and at least one excellent study has shown that the risk of eczema in babies can be reduced if they are given *Lactobacillus GG*. Chances are that this will work for some allergies as well. Particularly noteworthy is the fact that in over 150 studies of probiotics, no adverse effects have been noted.

Now the vexing question is to determine which probiotic bacteria are the most beneficial. *Lactobacillus GG* (modestly named by its discoverers, Sherwood Gorbach and Barry Goldin) looks very promising. It performs well against diarrhea, shows anticancer effects in animals, and in some cases has even relieved the symptoms of ulcerative colitis. So has VSL#3, a research mix of eight bacterial species. Bio-K+ is a commercially available product that in clinical studies has been shown to deliver the goods to the colon — namely, viable organisms in sufficient numbers. Still, there are products out there that claim to contain a host of beneficial bacteria but in fact do not. Let the buyer beware!

And what about plain old yogurt? If it is to be of any benefit, it has to have live cultures of acidophilus or bifido bacteria — preferably both. Check the label. Isn't it fascinating that the eccentric Dr. Kellogg may have been on the right track after all with his ideas about introducing beneficial bacteria into the colon? I'm taking his advice. I eat low-fat yogurt regularly. But as far as the enema machine goes, I'm just happy to have seen it in the museum.

Professor Wonder and Nutraceuticals

I don't think I would want to take nutritional advice from Professor Wonder. He seems to get things muddled. But maybe that's because he's not a real professor — he just plays one on TV. The good prof is the television spokesperson for the makers of that spongy, tasteless, presliced loaf that adorns many an American dinner table: Wonder Bread. Since many other breads vie for the same market, the people who make Wonder Bread have taken a shot at pulling ahead of the competition by enriching their product with calcium.

We all know how important an adequate calcium intake is for proper bone formation. Certainly, the dairy industry won't let us forget it. But when the makers of Wonder Bread added calcium to their recipe, the amount was too small to make a substantial contribution to a person's daily intake, at least as far as warding off osteoporosis goes. Perhaps this is why Professor Wonder decided to push the envelope a little and make the claim, "As a good source of calcium, Wonder Bread helps children's minds work better and helps their memory." Sounds good. There was just one little problem: he lacked scientific evidence to back up his claim. The public didn't question the professor's nutritional expertise, but the U.S. Federal Trade Commission did. It charged that the makers of Wonder Bread had made unsubstantiated health claims and violated federal law. The company agreed to refrain from making further claims unless they could be backed up by scientific evidence.

Of course, Wonder Bread is not the only manufacturer looking to cross the line between selling a food and selling a "nutraceutical." Simply put, a nutraceutical is a food or beverage that provides some health benefit beyond simple nutrition. The addition of iodine to salt, iron to cereal, and folic acid to flour are common examples of processes that were introduced to improve the health of consumers. A nutraceutical does not necessarily have to be a product of modern technology. Yogurt, for example, fits the bill. The live bifido and acidophilus bacteria present in many varieties may have beneficial effects. As we saw in the last chapter, these bacteria may elbow out some of the disease-causing microbes that invade the colon. They can be effective against certain types of diarrhea and may also help with bowel disorders such as Crohn's disease and irritable bowel syndrome. Furthermore, these bacteria may give a boost to the immune system and even reduce the risk of allergies if given to children at an early age. There is some question, however, about whether the helpful bacteria can survive the journey from the mouth, through the stomach and the small intestine, all the way to the colon. There may be another way to address the problem. A nutraceutical way.

I suspect that some people would raise an eyebrow if they encountered terms such as "fructooligosaccharides (FOS)," "lactulose," or "inulin" on a food label. These substances may not sound too appetizing, and you may not want to eat them, but the good bacteria in your large intestine find them to be tasty morsels indeed. As far as our bodies are concerned, these complex carbohydrates are just "fiber," meaning that we do not digest them as food. Our small intestines are not equipped with the enzymes they need to break down these carbohydrate polymers into smaller molecules that can be absorbed into the bloodstream. They therefore pass through the stomach and small intestine unchanged, and they collect in the colon. Here they help make up the bulk of the stool, and, more importantly, they foster the growth of beneficial bacteria and limit the multiplication of harmful ones. We refer to substances that stimulate the growth of specific bacteria in the colon as "prebiotics," and they are at the forefront of nutritional research. That's because the potential benefits include the prevention of abnormal cell proliferation (which can lead to cancer), improved mineral absorption, and even reduced blood cholesterol.

In Japan, numerous foods fortified with fructooligo-saccharides and inulin are already on the market, and the trend is coming our way. Where do these chemicals come from? They occur naturally in onions, garlic, and bananas, but not to an extent that would have a significant effect on colonic bacterial populations. We need a daily dose of at least four grams of prebiotics to have any hope of benefit, but double that amount is preferable. Just about the only way we can achieve such an intake is by adding fructooligosaccharides or inulin to processed foods. The most common source of the chemicals is chicory root, from which they can be readily extracted.

One plant that does contain a significant amount of inulin is the Jerusalem artichoke. The explorer Samuel de Champlain learned about this tuber from the North American Indians and introduced it to Europe. Actually, it is not an artichoke, and it has nothing to do with Jerusalem. The plant is a member of the sunflower family and is sometimes called a "sunchoke." But it seems that to Champlain it tasted like an artichoke, and the term stuck. Why Jerusalem? The Italians dubbed the new plant from America "girasole" — meaning "turning to the sun." Somehow this got corrupted to "Jerusalem."

In Europe and Japan, manufacturers add Jerusalem artichoke flour to foods to improve their health potential. It's kind of hard to find Jerusalem artichoke here, but I'm trying. Apparently, you can shred or slice the tubers into a salad or stir-fry.

There may be a downside. At least there is if we listen to John Goodyear, a British farmer who commented, in the 1860s, "In my judgment, which ever way they be drest and eaten, they stir up and cause a filthie loathsome stinking winde with the bodie, thereby causing the belly to be much pained and tormented, and are more fit for swine than for men." He may have been right about the wind, but he was surely wrong about the Jerusalem artichoke being unfit for human consumption. Indeed, I wouldn't be surprised if someday soon we hear Professor Wonder touting the benefits of Jerusalem artichoke flour, which he has added to Wonder Bread. He may even say something about improved calcium absorption. And you know what? The FTC would not object, because scientific evidence would back him up.

For Some, a Diet Goes against the Grain

Just ask people what they worry about most in their food supply and they'll round up the usual suspects. Their thoughts will drift to nitrites, sulfites, food colors, artificial sweeteners, monosodium glutamate, or genetically modified organisms. Yet we are far more likely to be harmed by a commonly occurring natural component in food than by any of these. Gluten, a protein found in wheat, barley, rye, and — to some extent — oats, can provoke health problems in a significant percentage of the population. Celiac disease, as gluten intolerance is usually called, may be much more common than we think.

Dr. Samuel Gee of Britain was the first to provide a clinical description of the disease. In 1888, he painted a disturbing picture of young children with bloated stomachs, chronic diarrhea, and stunted growth. Dr. Gee thought that the condition could

have a dietary connection, so he put his young patients, for some strange reason, on a regimen of oyster juice. This proved to be useless. Willem K. Dicke, a Dutch physician, finally got onto the right track when he made an astute observation during World War II. The German army had tried to starve the Dutch into submission by blocking shipments of food to Holland, including wheat. Potatoes and locally grown vegetables became staples, even among hospital patients, and Dicke noted that his celiac patients improved dramatically. Moreover, in the absence of wheat and grain flours, no new cases of celiac occurred.

By 1950, Dicke had figured out what was going on. Gluten, a water-insoluble protein found in wheat, was the problem. As later research showed, the immune systems of celiac patients mistake a particular component of gluten, namely gliadin, for a dangerous invader, and they mount an antibody attack against it. This triggers the release of molecules called cytokines, which in turn wreak havoc upon the villi — the tiny, fingerlike projections that line the surface of the small intestine. The villi are critical in providing the large surface area needed for the absorption of nutrients from the intestine into the bloodstream.

In celiac disease, the villi become inflamed and markedly shortened, and their rate of nutrient absorption is effectively reduced. This has several consequences. Nonabsorbed food components have to be eliminated, and this often results in diarrhea. Bloating can also occur when bacteria in the gut metabolize some of these components and produce gas. But the greatest worry for the celiac disease sufferer is loss of nutrients. Protein, fat, iron, calcium, and vitamin absorption can drop dramatically, and this results in weight loss and a plethora of complications. Luckily, if the disease is recognized and a gluten-free diet initiated, the patient can lead a normal life.

Diagnosis of celiac disease involves the physician taking a biopsy sample from the duodenum, the uppermost section of the small intestine, via a gastroscope passed down through the patient's mouth. Microscopic analysis shows the damaged villi. Recently, blood tests have also become available. One of these screens for the presence of antigliadin antibodies, but it is not foolproof. Only about half the patients with positive results actually show damaged villi upon biopsy. The antitissue trans-glutaminase test (anti-tTG) is a much better diagnostic tool, but it is available only in specialized labs.

There is a great deal of interest in these tests because of their potential value in identifying celiac cases and perhaps even in screening the population. Celiac disease, which has a genetic component, does not necessarily begin immediately after gluten is first introduced into the diet. The onset of disease can occur at any age. In adults, the symptoms are usually much less dramatic than they are in young children. The first signs are often unexplained weight loss and anemia due to poor iron and folic acid absorption. A patient's stools tend to be light-colored, smelly, and bulky because of unabsorbed fat. Symptoms can include a blister-like rash, joint and bone pain, stomachache, tingling sensations, and even headaches and dizziness. Identification of celiac patients is important not only because much of their misery can be prevented by a gluten-free diet, but also because a recent study showed that over a thirty-year period the death rate among celiac patients was double that expected in the general population. Delayed diagnosis and poor compliance with diet increase the risk. The major cause of death among celiac disease sufferers is non-Hodgkin's lymphoma, a type of cancer known to be associated with celiac disease. A less severe but more common complication than cancer is osteoporosis, brought on by poor absorption of calcium and vitamin D.

Unfortunately, a gluten-free diet is not that easy to follow. Wheat and barley crop up in a wide assortment of products. Patients have to become veritable sleuths; they quickly discover that foods as diverse as ice cream, luncheon meats, ketchup, chocolate, and even communion wafers can contain gluten. Fortunately, the Celiac Association distributes excellent information on dietary dos and don'ts, and consumers can now purchase a large assortment of gluten-free products based on rice, corn, and soy.

The plan of action for biopsy-diagnosed celiacs is clear. They must adhere religiously to a gluten-free diet in order to eliminate symptoms and reduce the risk of osteoporosis and lymphoma. But what about people who have no overt symptoms yet have a positive blood-test result? Surveys indicate that one in about two hundred people may fall into this category. The biopsies of some may show normal villi initially, but these people are considered to have latent celiac disease, which may become symptomatic years later. Others may have flat villi without symptoms, a condition referred to as silent celiac disease, which can become aggressive at any time. Should these people follow a preventative, difficult-to-maintain diet? At this point, nobody really knows, since we still have much to learned about the effects of gluten. Recently, for example, researchers discovered that celiac patients who complained of headaches showed brain inflammation on MRI scans, and that the problem resolved on a gluten-free diet. Certain individuals have provided anecdotal and controversial evidence that the condition of some autistic children improves when gluten is eliminated from their diets. However, there is no evidence that these children have celiac disease.

So, on the one hand, it certainly seems that we have not yet uncovered all of gluten's potential mischief. On the other hand, one intriguing method of reducing gluten exposure has emerged. Preliminary research suggests that it may be possible to remove gluten's offensive component by genetically modifying wheat. That would be a boon to celiac patients and perhaps even to those who may be suffering in silence.

SACCHARIN: FROM BACK ALLEY TO TABLETOP

"Psst . . . I've got some of the good stuff here," the shadowy figure in the back alley whispered. Word spread quickly, and soon people were scurrying from everywhere to purchase little bags of the white powder. They tasted it carefully, hoping for just the right sensation. But it was not euphoria they were after, it was sweetness. All they were looking for was a personal supply of contraband saccharin. Then life would be sweet once more.

In 1902, the German government passed legislation prohibiting the sale of saccharin to healthy consumers. Pharmacies could still sell the sweetener to diabetics, but others would have to make their strudel the old-fashioned way — with sugar. If they could afford it. The saccharin ban had nothing to do with safety concerns; it was the result of lobbying efforts by the powerful beet-sugar industry. Saccharin was cheap to produce and had quickly become the "sugar of the poor." Fearing the loss of tax revenue from the sugar manufacturers, many European governments agreed to curb the sale of saccharin. This gave rise to a huge black market, amply supplied by producers in Switzerland, where saccharin remained legal. It wasn't until World War I that a sugar shortage brought saccharin back into favor in Europe. It rapidly became the prime sugar substitute, a status it has retained to this day. But most people who sweeten their lives with saccharin have no idea of the fascinating story that lies behind its discovery.

It all started the day a young American named Ira Remsen noticed a bottle sitting on a table in his doctor's office. Most people would not have given it a second glance, but to Ira, that bottle of nitric acid presented an unexpected opportunity to clear up the meaning of a confusing phrase he had come across in his chemistry book. What did "nitric acid acts upon copper" really mean? Here was his chance to find out. Digging in his pockets, he found a couple of copper pennies. Working quickly, Ira poured a little of the acid over the coins, and, to his amazement, this produced an immediate reaction. "A green-blue liquid foamed and fumed over the cent," he later recalled. "The air in the neighborhood of the performance became dark red. A great colored cloud arose. This was disagreeable and suffocating!"

When Ira tried to fling the mess out the window, he discovered that nitric acid acts not only upon copper, but also upon fingers. Wiping his burning fingers on his trousers, he made another discovery: nitric acid also acts upon fabrics. And so it was that Ira Remsen learned the meaning of the common chemical expression "acts upon." But he learned something else as well. He learned that the only way to understand chemical action is to experiment and witness the results — in other words, to work in a laboratory. On that day back in the 1860s, Ira Remsen decided to dedicate his life to the fledgling field of chemistry.

However, a career in chemistry was not what Ira's parents had in mind for their son. They wanted him to be a doctor. So Ira dutifully worked for, and achieved, a doctorate in medicine. But as soon as he'd done that, he left his home in the United States and went to Germany to pursue his real love. With a doctorate in chemistry in hand, he returned to the U.S. to take up a position as professor of chemistry at Williams College. The job was a frustrating one, because, unlike their German counterparts, American colleges did not make research a priority. Remsen was forced to concentrate on teaching, but he taught with great enthusiasm. He resolved that his students would experience and understand chemical reactions, starting, of course, with the one between nitric acid and copper. The reddish vapor that filled the air was due to the formation of nitrogen dioxide, he explained, and the bluish residue was copper nitrate. Remsen also wrote a widely acclaimed textbook, which introduced students to the principles of chemistry. His increasing renown won him the chemistry chair at a new university that had been founded in Baltimore with an endowment from a local merchant by the name of Johns Hopkins.

Remsen jumped at the opportunity to get in on the ground floor and establish a program that would focus on research. Indeed, Johns Hopkins University soon became the center for the study of chemistry in the United States, attracting students and researchers from around the world. One of these was Constantine Fahlberg, a German chemist who wanted to continue his studies under Remsen. The project Remsen assigned him was not a particularly exciting one. This wasn't surprising, since Remsen was interested in science as a form of higher culture, not as a practical tool to solve problems. He asked Fahlberg to study the oxidation of certain coal-tar derivatives known as toluene sulfonamides, simply because no one had done this before. Fahlberg, it seems, was a pretty sloppy chemist — he often didn't bother to wash his hands before leaving the laboratory. His sloppiness, though, turned into a stroke of luck.

At dinner one evening, Fahlberg noticed that the slice of bread he had picked up tasted unusually sweet. It didn't take him long to figure out what had happened. He traced the sweetness to a substance he had been handling in the laboratory, and he immediately brought this chance discovery to the attention of Remsen. In 1880, the two scientists published their finding in *The American Chemical Journal*, noting that the new compound was hundreds of times sweeter than sugar. Remsen looked upon this as a mere curiosity, but Fahlberg immediately saw the potential for commercial exploitation. He knew that sugar prices fluctuated greatly, and that a low-cost sweetening agent would be most welcome. Those on a weight-loss regime, Fahlberg thought, would also find the new product appealing. The product would dramatically lower the calorie count of sugar-sweetened foods, since it was so sweet that only a tiny amount would produce the desired sweetness. Fahlberg coined the term "saccharin" for his discovery, after the Latin word for sugar, and he secretly patented the process for making it. Within a few years, saccharin became the world's first commercial nonnutritive sweetener, and it made Fahlberg a wealthy man.

Remsen did not resent the fact that neither he nor Johns Hopkins University ever made a dime from saccharin. He was a pure scientist at heart and did not much care whether his research turned out to be financially profitable. But he did develop an intense dislike for Fahlberg, who, by all accounts, tried to take sole credit for the discovery. "Fahlberg is a scoundrel," Remsen often said, "and it nauseates me to hear my name mentioned in the same breath with him!" But due to the importance of the saccharin discovery, their names will be forever linked. The importance is twofold: first, the commercial production of saccharin is the earliest example of a technology transfer from university research to the marketplace; second, and more importantly, saccharin introduced the concept of a non-nutritive sweetener, an idea that has been mired in controversy from the moment it was first raised.

Saccharin went into commercial production in Germany, where Fahlberg had taken out a patent. It wasn't until 1902 that John Francis Queeny, a former purchasing agent for a drug company in St. Louis, decided to take a chance on manufacturing saccharin in the United States. Here the sweetener was not burdened by any of the legal problems that were arising in Europe. He borrowed fifteen hundred dollars and founded a company that at first had only two employees — himself and his wife. Queeny decided to give the company his wife's maiden name, and Monsanto was born. At first, the company's only product was saccharin, but it quickly diversified to become one of the largest chemical companies in the world.

The unfettered use of saccharin in America did not last long, however — thanks mostly to the work of Dr. Harvey W. Wiley, who, in 1883, was made chief of the Department of Agriculture's Bureau of Chemistry. The bureau had been created to monitor the safety of the food supply when the population began to increase dramatically after the Civil War, resulting in large-scale changes in food production. People were moving into the cities from farms, and they no longer ate every meal at home or prepared every meal from scratch. A burgeoning food industry was gearing up to meet the demand for prepared foods and the preservatives needed to make them safe. Wiley had become concerned about the unregulated use of such food additives, an issue to which he had become sensitized during his days as a professor of chemistry at Purdue University. In 1881, he had published a paper on the adulteration of sugar with glucose, and he had looked into the problem of coloring cheese with lead salts. Now Wiley worried about the extensive use of formaldehyde, benzoic acid, and boric acid as preservatives. Since they poisoned bacteria and molds, could they also poison humans? He decided to find out.

The hallmark of Wiley's crusade for safer food was the establishment of the celebrated Poison Squad. Wiley recruited twelve healthy young men, asking them to meet every day for lunch and dine on foods prepared with a variety of additives. If the men developed any unusual symptoms, Wiley would move for a ban of the additive. In retrospect, this was a primitive system, because it revealed nothing about exposure to small amounts of chemicals over the long term. Still, Wiley's work publicized the need for food regulation, and his efforts finally culminated in the passage of the Pure Food and Drugs Act of 1906, which for the first time gave the government some teeth with which to bite food and drug adulterers.

Dr. Wiley became a food-safety zealot, and he caught saccharin up in the net he cast to catch chemical culprits. He vigorously attacked saccharin as a "coal-tar by-product totally devoid of food value and extremely injurious to health." Unfortunately for Wiley, President Theodore Roosevelt had been prescribed the sweetener by his physician, and he loved the stuff. "Anyone who says saccharin is injurious to health is an idiot," Roosevelt proclaimed, and he decided to curtail Wiley's authority. The president established what he called a "referee board of scientists" — ironically, with Ira Remsen as its head — to scrutinize Wiley's recommendations. The board found saccharin to be safe but suggested that its use be limited to easing the hardship of diabetics. That suggestion had no legal bearing, and it was soon forgotten in the face of massive industry maneuvering to satisfy the public's demand for nonnutritive sweeteners.

The saccharin bandwagon rolled happily along until 1977, when a Canadian study suggested an increased incidence of bladder cancer in male rats fed the equivalent of eight hundred diet drinks a day — male rats whose mothers had been dosed with the same amount of saccharin. Though the study was ridiculed by saccharin promoters as irrelevant to human subjects, the Canadian government banned saccharin as a food additive but allowed its continued use as a sweetener that consumers added themselves. The U.S. Food and Drug Administration (the descendant of Wiley's bureau) also proposed a ban, but a massive public outcry prompted Congress to permit the sale of saccharin pending further studies. Its continued use as an additive was allowed, but a warning label on the familiar little pink packets stating that saccharin was known to cause cancer in laboratory animals was mandated.

Subsequent research failed to clear saccharin of all blame as a carcinogen, but human epidemiological studies have shown that if there is any risk at all, it is a very small one. In fact, in 2000 the U.S. government removed saccharin from its official list of human carcinogens, and President Clinton signed a bill eliminating the requirement for a warning label on the product. Canada still does not allow saccharin as an additive. But you don't have to purchase it stealthily in back alleys. You can buy it legally in pharmacies. Diabetics are certainly grateful for that.

Ira Remsen, I'm sure, could never have imagined where his little experiment with nitric acid would eventually lead. And why was there nitric acid in the doctor's office in the first place? Because in those days, silver nitrate was used as an antiseptic and was generated by nitric acid "acting upon" silver.

Aspartame: Guilty or Innocent?

There is an important aspartame story to be told. I'm just not sure what it is. Is this artificial sweetener the chemical from hell that is responsible for causing multiple sclerosis, brain tumors, seizures, and lupus, along with a host of other ailments, or is it one of the safest and best-tested food additives on the market? The answer seems to depend upon whom you ask.

We had best begin by covering the available facts about aspartame. The sweetener is commonly labeled "noncaloric," although that term is technically inaccurate. Aspartame is broken down in the digestive tract into its components — namely, aspartic acid, phenylalanine, and methanol — which the body absorbs and metabolizes. Collectively, they contribute about four calories per gram, but since the substance is about 180 times sweeter than sugar, very little needs to be used in foods and beverages to achieve a satisfactory degree of sweetness. So the calorie contribution is essentially irrelevant.

Diet drinks normally contain about sixty milligrams of aspartame per one hundred milliliters, which translates into roughly two hundred milligrams per serving. To put this into perspective, we need to introduce the concept of acceptable daily intake (ADI), which the U.S. Food and Drug Administration uses to describe an intake level that, if maintained each day throughout a person's lifetime, would be considered safe. The ADI for aspartame is fifty milligrams per kilo of body weight. The actual average daily intake is less than two percent of this, and even the heaviest aspartame consumer ingests only about sixteen percent of the ADI. To reach the ADI, an adult would have to drink twenty twelve-ounce soft drinks; a child would have to drink seven. An adult would have to consume ninety-seven packets of sweetener. Industry figures show that ninety-nine percent of aspartame users consume less than thirty-four milligrams per kilo of body weight per day. The average consumption is about five hundred milligrams per day. The ADI for a person weighing seventy kilos (about 150 pounds) is 3,500 milligrams. Aspartame cannot be used in cooked or baked foods since it breaks down into its components upon exposure to heat and loses its sweetening power.

Aspartame is perhaps the most widely researched food additive ever to have landed on the market. Its manufacturers were expecting reports of adverse reactions, because that is so often the case with newly introduced substances (and no amount of testing can preclude idiosyncratic reactions in some portion of the population); in fact, there have been only a few such reports about aspartame. Over seventy million North Americans consume aspartame on a regular basis, yet reported complaints average only about three hundred per year. The majority of complaints (sixty-seven percent) refer to headaches, dizziness, visual impairment, and mood alterations. Aspartame consumers have also reported gastrointestinal problems (twenty-four percent) and allergic symptoms such as hives, rashes, and swelling of tissues (fifteen percent). Some have linked seizures with aspartame exposure. In most instances, these difficulties occurred when aspartame intake far exceeded normal use.

Researchers have conducted double-blind challenges with aspartame. At Duke University, in one of the best designed of such studies, they investigated the effects of a single large dose of aspartame in people who had claimed to be sensitive to the substance. The results showed no difference in headache frequency, blood pressure, or blood histamine concentrations (a measure of allergenic potential) between the experimental groups and the control groups.

In another study, carried out at the University of Illinois and involving diabetics, subjects in the placebo group actually had more reactions than those in the aspartame group. Still, surveys taken by physicians in headache clinics reveal that aspartame precipitates headaches about eight percent of the time. This kind of conflicting data is characteristic of the research on the possible side effects of aspartame. Reported anecdotal experiences are not confirmed by carefully controlled scientific studies. This does not mean that the problems aren't real, but it does imply that in many cases the symptoms are not caused by aspartame. People get headaches, upset stomachs, aches and pains of all kinds on a regular basis for no easily determined reason. If they recall having consumed aspartame just before one of these ailments struck, they may judge the sweetener guilty by association. This is even more likely if they are already familiar with some of the bad publicity that aspartame has received.

Sometimes a reaction attributed to aspartame may be due to something else that we can pinpoint. A caller on my radio show informed me with great conviction that the palpitations she had been suffering from for years disappeared immediately after she gave up aspartame-sweetened drinks. This seemed unlikely to me, so I asked her what she thought would happen if she drank a regular cola. "Nothing!" was her immediate reply. She agreed to take this simple test, and she was shocked when the palpitations resumed. The lady had been reacting adversely not to aspartame, but to caffeine, yet she had been convinced that aspartame was the culprit.

Certainly, there have been studies that have found adverse reactions to aspartame. At least one has confirmed allergic symptoms such as hives and swelling in sensitive individuals. We don't really know how the allergy comes about, since researchers believe that none of the components of aspartame produces allergic reactions. But some have suggested that diketopiperazine, a compound that forms when aspartame decomposes, may be responsible.

Without a doubt, the three breakdown products of aspartame are all toxic in high doses. Phenylalanine is an essential amino acid, which must be included in the diet for normal growth and tissue maintenance, but sustained high blood levels of the substance can lead to brain damage. This is of major concern to the roughly one out of twenty thousand children who are born with an inherited condition called phenylketonuria, or PKU. These children cannot metabolize phenylalanine properly, and the substance builds up to dangerous levels in their brains. The condition therefore necessitates a severe curtailment of phenylalanine intake, at least for the first six years of life. This means that aspartame, due to its phenylalanine content, is not suitable for PKU sufferers; those who manufacture products containing aspartame must therefore include a warning on their products' labels.

In members of the general population, blood levels of phenylalanine after aspartame ingestion fall into the same range as they do after ingesting any protein-containing food. Even at abusive amounts — equivalent to a child swallowing one hundred sweetener tablets — levels do not rise above those considered to be safe in children afflicted with PKU. The effects of aspartic acid, another aspartame breakdown product, have also been rigorously examined. When researchers administered extremely large amounts to primates, no damage resulted, even though the subjects' blood levels were greatly elevated. Humans eliminate even high doses very quickly. Most significantly, aspartic acid levels in our blood do not rise after we eat foods containing aspartame, or after we drink sweetened beverages, even at the rate of three drinks in four hours.

Perhaps the most curious accusations leveled against aspartame have involved its methanol content. It is a fact that in large doses methanol can cause blindness, and even death. Alarmists have therefore referred to the methanol that is released from aspartame as an unsafe substance. We must remember, however, that there are no safe substances, only safe doses. The amount of methanol released from aspartame is inconsequential when we look at it in the context of the overall diet. Methanol occurs naturally in foods. The "natural" methanol content of fruit juice is actually about two and a half times greater than the quantity of methanol released in the body from a diet drink sweetened with aspartame. Aspartame opponents argue that our bodies handle methanol differently when we ingest it along with other alcohols, such as ethanol, which are found in juices. There is no evidence for this; and, furthermore, even the blood of those subjects who consume the greatest amounts of aspartame contains no detectable quantity of methanol.

One study, which was published in 1996, claimed that a ten-percent increase in brain tumors noted in the 1980s was associated with the introduction of aspartame. This study received a great deal of publicity. It suggested that aspartame, or its diketopiperazine breakdown product, may combine with nitrites in the diet to form nitrosated compounds. These compounds are indeed known to produce brain tumors in animals, but the manufacturers of aspartame point out that while aspartame use has increased dramatically since the 1980s, brain-tumor rates have not increase.

If we look at aspartame from a scientific perspective, we can hardly see it as a dastardly poison. But one could certainly get a different impression by reading a letter, attributed to one Nancy Markle, that has been making the cyber rounds. It alleges that virtually every ailment known to humankind is caused by aspartame. I talked about this issue on my radio show one day, commenting that such arguments are not supported by the vast majority of studies published in the scientific literature. It was at this point that I received an e-mail from Betty Martini, who runs Mission Possible International, an organization dedicated to ridding the world of the deadly scourge of aspartame. I was too soft on aspartame, she insisted, and I should open my eyes and see the misery that this devil of a chemical is inflicting on the public. Ms. Martini has since flooded me with information on aspartame — particularly, dozens of anecdotal reports from people who claim that their lives, or the lives of their loved ones, were ruined by aspartame. She also included accounts of the miraculous turnarounds people experienced when they gave up the vile substance. Without any evidence to back her up, the lady also maintained that thousands of Gulf War veterans succumbed to aspartame disease because they consumed diet drinks that were not refrigerated. Methanol, she explained, is released at warm temperatures, and it is the culprit.

Betty Martini is absolutely convinced that she has discovered a major cause of human suffering, and she has undertaken what she feels is an essential crusade against it. But the facts simply do not justify an all-out war on aspartame. The human body and the human mind are extremely complex. The likelihood that a single chemical entity can cause such a wide diversity of problems is small, but we often have a need to see cause-and-effect relationships where none exist. Numerous publications and Web sites promote dietary supplements, electrical devices, and specialized waters with testimonials galore. Other sites are dedicated to ridding the world of evils ranging from soy protein and canola oil to antiperspirants and dental amalgam. The allegations made by the contributors to these sites parallel those made by the foes of aspartame. But can there really be so many different causes and cures for the same conditions?

I am not a fan of aspartame, or any other artificial sweetener, although I recognize that they have certain advantages for diabetics. (Nondiabetics, however, should develop healthy diets instead of seeking shortcuts such as artificial fats or sweeteners.)

But aspartame was not cast onto the market in a thoughtless fashion. Researchers conducted hundreds of studies on its safety profile. Yes, there are some problems. As I pointed out earlier, in a rare documented case aspartame has been linked to seizures. Some aspartame consumers have experienced headaches and episodes of dizziness. And one small study did show that people suffering from depression found that their condition deteriorated when they consumed the artificial sweetener. Yet these problems occurred only with high doses — and, as we well know, toxicity is related to dosage. Unfortunately, many people do consume staggering amounts of artificially sweetened products, starting their day with a six-pack of diet soda and finishing it with an array of diet desserts. This is an abominable practice, and we should heartily discourage it. Drinking a single diet beverage or chewing a stick of sugarless gum, however, is quite another matter.

We scientists need to look at all the available research. If we opt for selective viewing and blind dedication to various agendas, we will delude ourselves. Is multiple sclerosis, or MS, linked to aspartame consumption? The vast majority of researchers say no. A Medline search reveals no links to MS or lupus. Brain cancer? Dr. John Olney of Washington University maintains that there is a link. He claims that an increase in brain tumors has paralleled aspartame's rise in popularity. Other researchers have demonstrated that this increase began about eight years before aspartame was introduced and that it has now leveled off, despite the fact that aspartame use has skyrocketed. Betty Martini's answer to this divergence of opinion is that the MS associations, the FDA, and many researchers have been bought off by the manufacturers of the artificial sweetener. She would probably say the same thing about the researchers in California who published a paper in a 1997 issue of *The Journal of the National Cancer Institute* describing how they collected information on aspartame exposure from fifty-six brain cancer patients, all under age nineteen, and compared their intake to that of ninety-four controls. Patients with tumors were no more likely to have consumed aspartame, and maternal consumption did not elevate risk in either breast-fed or bottle-fed children.

So, who is right? Well, I thought I would clear the air by going to the top. And at the top of this field is Dr. Richard Wurtman of MIT, probably the world's leading expert on the relationship between nutrition and the brain. Betty Martini often refers to his work, and she calls Dr. Wurtman a brilliant researcher. He does not return the compliment. I was fortunate enough to have a long telephone conversation with Dr. Wurtman, and he told me that while he believes that, in rare cases, large amounts of aspartame may cause seizures, he sees no problems with the amounts people normally consume. He consumes diet drinks himself.

What is perhaps the best double-blind study ever carried out in this area failed to find an aspartame effect. Dr. Paul Spiers and some of his colleagues at MIT gave subjects aspartame at a dose equivalent to more than a dozen diet drinks a day, and they found no difference in brain waves, mood, memory, behavior, or physiology. Reports of headache, fatigue, and nausea occurred with equal frequency in the aspartame and the placebo groups. Opponents of this research cast a shadow on the study, saying that it was supported financially by the artificial-sweetener industry. But where else should researchers seek funds for sweetener research? From a lightbulb manufacturer? The fact that a researcher receives a grant does not indicate that he or she has been bought off.

Not all of the e-mail I received on this issue was of a scientific nature. Alex Constantine, who writes prolifically on aspartame, became involved in the correspondence between me and Betty

Martini, and he offered up this bit of wisdom: "Joe Blow [he means me] works at McGill University, once the most active hive of mind-control experimentation in the world and still very involved with the CIA. He writes for *Reader's Digest*, a CIA publication. A professional prostitute on the CIA payroll. A fascist collaborator who smears antifascists for fun and profit."

I will certainly continue to monitor the aspartame story closely. The fact that so many people believe it to be harmful and claim benefits when they give it up is interesting, albeit anecdotal. Science, however, often starts with anecdotal evidence. But this must progress to proper, controlled studies. And while, without a doubt, some people do experience adverse effects, these studies, to date, do not support the allegations of a worldwide epidemic of "aspartame disease." Let's allow the facts to speak. As Mark Twain quipped, "The worst kind of ignorance is the things we know for sure that just ain't so."

An Ode to the Oat

I'd like to take a look at Papa Bear's blood test. His triglycerides are probably high from slurping all that honey, but his cholesterol level is likely to be just fine, thanks to his love of porridge (ref: Goldilocks). In fact, all the members of the Bear family, with their penchant for oats, can serve as nutritional role models. I, for one, am following in their footsteps. And I'm managing to keep pace with science.

The Scots got this one right. Porridge is one of their staples. Scotch oats are steeped not only in water and milk, but also in a good dose of tradition. I understand that the mush must be stirred clockwise, with the right hand, using a "spurtle," which is a sort of wooden stick especially made for this purpose. And the porridge is to be eaten from a birch-wood bowl. "Porridge sticks to the stomach and scrubs out the bowels," the Scots maintain. True enough. Oats really do have a high satiety value. Essentially, this means that they take a long time to digest and therefore keep you feeling full longer. Indeed, in a study comparing oatmeal to cornflakes as breakfast foods, researchers found that subjects who ate oatmeal consumed one-third fewer calories for lunch. So, oats can help you lose weight.

The bowel-scrubbing bit makes sense too. In more ways than one. Oats contain fiber. Fiber is the structural part of plants, grains, fruits, and vegetables; it cannot be broken down by enzymes in our digestive tract and therefore cannot provide nutrition. In other words, most of what you eat turns into you, but fiber passes through. There are two kinds of fiber: insoluble and soluble. Cellulose is the classic insoluble fiber, whereas pectin, found in fruits, is an example of the soluble variety. The former keeps us regular, reduces the risk of diverticulitis, and helps eliminate substances that may play a role in colon cancer. But it is beta glucan, the soluble fiber in oats, that is causing a stir. Solid research has shown that while oats produce no nutritional miracles (no single food does), those who consume them regularly can experience these benefits: lower blood cholesterol levels, a decrease in high blood pressure, healthy arteries, and better diabetes control.

Some of this information about oats is not new. Just think back to the oat bran craze of a few years ago. Retailers couldn't keep the stuff on the shelves. Rumors of a new shipment sent anxious shoppers rushing to the supermarket, only to have their hopes dashed when they found that the booty had already been snapped up. Why was there such a feverish interest in a product traditionally considered animal feed, not human food? Because some tantalizing studies showed that oat bran, the outer covering of the grain, is an excellent source of soluble fiber, which has the ability to reduce cholesterol. Some researchers offered a theory to explain how this happens. Beta glucan absorbs water in the intestine and forms a viscous slurry that traps cholesterol from food as well as some of the bile acids needed for digestion. Since these compounds are made in the body from cholesterol, their removal from the digestive tract forces the body to synthesize more. The result is a depletion of the cholesterol in the blood. Good stuff. But there was a problem. The public never got the proper message about how much oat bran they would have to consume to create an impact on their blood cholesterol levels. And this was no small amount.

To reduce blood cholesterol by roughly five percent, a person needs to eat three to four grams of beta glucan a day. More is not better. At higher doses, one experiences a sense of fullness; bloating and gas production become apparent. Now, a five-percent reduction doesn't sound like a lot, but it can lower the risk of a heart attack by as much as ten percent. We can find this amount of beta glucan in one cup of cooked oat bran, or one and a half cups of oatmeal. Three packets of instant oatmeal will do it too. But oat bran cookies, oat bran chips, and oat bran gum will not. Yet manufacturers flooded the market with these silly products, hoping to capitalize on the oat bran mania. The products had no effect on cholesterol, and they tasted lousy to boot. Little wonder the oat bran fad faded quickly. Too bad. Because, when consumed in the right quantities, oats really do deliver the goods. They can do more than just lower cholesterol — they can reduce blood pressure.

A pilot study in Minnesota focused on a group of patients who took at least one medication for hypertension. Researchers asked half of them to consume about five grams of soluble fiber per day in the form of one and a half cups of oatmeal and an Oat Square (an oat-based snack); they asked the other half to eat cereal and snacks with little soluble fiber. Oat consumption reduced blood pressure in these patients significantly. Indeed,

about fifty percent of them were able to give up their medication. How oats lower blood pressure is not clear, but it probably has to do with modifying insulin response. The pancreas secretes insulin, which enables our cells to absorb glucose from the bloodstream after a meal. A glucose surge triggers a quick insulin response, but if such surges are frequent, insulin becomes less effective, and the body needs to produce more and more. This leads to a condition known as insulin resistance. Researchers suspect that such insulin resistance plays a significant role in elevating blood pressure by constricting blood vessels. Soluble fiber slows the absorption of nutrients from the gut and blunts the insulin response. This also explains why oats can help diabetics control their blood sugar levels.

And if that weren't enough to boost your appetite for oats, just consider that oats contain a unique blend of antioxidants, including the avenanthramides, which prevent LDL cholesterol (the bad cholesterol) from being converted to the oxidized form that damages arteries. So, it isn't hard to see why I've become a real oat fan. And I've become an even bigger fan since I discovered "steel-cut oats." These are oat grains cut into thirds but not rolled into little flakes by a machine. They have a great nutty flavor. Admittedly, they do take longer to cook and require constant stirring. That's why I'm searching for a good spurtle. If you've got one, I'll trade you for my oat soup recipe.

Ah, heck. I'll give you the recipe anyway. Bring twelve cups of chicken stock to a boil. Add six sliced carrots, three sliced parsley roots, one cup of peas, one cup of diced onion, two tablespoons of canola oil, two tablespoons of soy sauce, two mashed garlic cloves, and two cups of rolled oats. Simmer for forty minutes and add salt and pepper to taste. I bet even Baby Bear would love it.

THE SECRET LIFE OF BAGELS

You should have seen the face of the guy behind the counter in the Manhattan bagel shop when I asked for the smallest, thinnest bagel they had. In a country where excess rules, where the credo is "bigger is better," my request must have come as a shock. But I really needed that thin bagel to save a lecture I was about to give at Columbia University.

The focus of my lecture was on some interesting everyday applications of chemistry, and I wanted to start with a demonstration of how acrylic plastics can make our lives less risky. Dr. Mark Smith, head of emergency at George Washington University Medical Center, had made headlines across America by going public about a "great underreported injury of our times": cuts resulting from bagel slicing. Anyone who has ever risked a mangled hand by trying to slice a bagel in half knows exactly what I'm talking about. Luckily, inventors have risen to the challenge and have come up with a variety of devices to ensure that a perfectly good bagel isn't ruined by splattered blood. I had even found one that I really liked. It was a clear acrylic box that held a bagel snugly and had slits down two sides to guide a knife. Not only does it prevent injuries, it also protects bagel lovers from another great scourge — a smoke-filled kitchen. This is what happens when the bigger half of an unevenly sliced bagel refuses to pop up after we've squeezed it into a toaster slot that is too small.

My proposed demonstration of scientific bagel cutting obviously required a victim, and I planned to order that victim at breakfast. Alas, what they brought me was a gigantic roll with a hole in it that looked more like a life preserver than a bagel. I realized that I had a problem. There was no way this thing would fit into my bagel cutter. That's when I ran to the bagel shop and made my unusual request. No shortage of bagels here, but all were as obese as my original. And then, as I stood there frustrated, the door to the back of the shop flew open, and I caught a glimpse of what was going on. Employees were sending raw bagels through a steaming machine. They weren't boiling them, they were steaming them. That's when I decided that New Yorkers didn't need to learn about acrylics. They needed to learn about bagel making.

Montreal is the center of the bagel world, because here we do it right. For just 180 calories and virtually no fat, you get splendid flavor, unique texture, and a dose of history. According to legend, in 1683, King John Soviesky of Poland helped save Vienna from Turkish invaders. A grateful Viennese baker created a stirrup-shaped roll to commemorate the bravery of the Polish soldiers. In a German dialect, this roll came to be called "beugel" — meaning "ring," or "bracelet" — because of the large hole in its middle. "Beygel" was the Yiddish version of the name, and from this it was only a short hop to "bagel." The bagel was

introduced to North America by Jewish immigrants about a century ago, and in Montreal some of their descendants are still delighting customers by producing bagels in the traditional fashion. There's nothing like the smell and taste of a fresh bagel straight out of the oven. Try the bagel challenge. I defy anyone to buy a dozen and still have a dozen by the time they arrive home. Cannot be done. Not even by someone reared on sliced white bread.

To make this gustatory and health marvel, you don't start with just any flour; you use a flour that is rich in two proteins: glutenin and gliadin. These long, coiled, tangled molecules unfold and line up in long strands when kneaded with water. They also forge cross-links with each other, building a network of proteins known as gluten, which gives dough the elasticity it needs to rise as yeast generates carbon dioxide gas. The baker adds a small amount of sugar to the dough to serve as food for the yeast, along with a little egg for color and flavor. Kneading is critical, because it creates the air pockets into which the carbon dioxide will expand. These air cells will contribute greatly to the final texture. Furthermore, oxygen in the air, introduced during kneading, strengthens the gluten by promoting a chemical reaction that forms sulfur-sulfur links between adjacent protein molecules.

What makes a bagel a bagel, however, is neither the flour nor the kneading. It is the immersion of the hand-formed rings of dough in boiling water prior to baking. Starch molecules in flour are coiled together in tiny granules, but hot water penetrates the granules and causes them to swell. Then the swollen granules muscle their way into, and strengthen, the molecular scaffolding created by the gluten proteins. A classic chewy bagel is the result. Furthermore, the boiling water is not just any boiling water. The baker must dissolve a little honey in it. That's because, in the heat of the oven, sugars in the honey combine with proteins in the dough to form the shiny brown crust prized by bagelites.

Ah, the oven. You can't make a proper bagel without a wood-burning oven. The smoke enhances the flavor, and the burning wood provides just the right temperature. During baking, gluten coagulates, and starch completes its gelatinization. If the temperature is too low, the dough will expand as the volume of the trapped gases increases, but it will then collapse because the gluten and starch have not set. If the oven is too hot, the setting takes place too soon, and the dough does not gain enough volume. It's a touchy business that needs an expert hand. A Montreal hand.

What I saw in New York was not a pretty sight. I saw dough being steamed instead of boiled. I saw electric ovens. I saw jalapeno peppers, chocolate chips, and — believe it or not — bacon bits added to bagels. But even this sacrilege did not prepare me for what I was to see in the frozen food section of the supermarket into which I dashed, hoping against hope, to find a bagel that looked like a bagel. Staring me in the face was the "UnHoley" bagel. It looked like a hamburger bun filled with cream cheese. No hole. No class.

By this time, I was getting desperate. I was frustrated by bagels that had no holes and others that were like king-sized doughnuts with rigor mortis. I had one last chance — Zabar's, Manhattan's most famous food store. No proper bagels here, either, but Zabar's did have something to save the day. An adjustable bagel cutter. It was polyethylene, not acrylic, but I just adjusted my talk accordingly. Thank goodness for American ingenuity.

A Toast to Toast

I sprinkled a little lysine into the gently boiling corn syrup. Within minutes, the kitchen was filled with the aroma of freshly toasted bread. Then I added a little of this brew to a cup of hot water and tasted it. Sure enough, it tasted like toast. Actually, it tasted almost exactly like another potion I sometimes prepared by soaking toasted bread in water. By now you're probably thinking that I must lead a pretty dull life if I have to resort to investigating the properties of toast water, but such is not the case. When the public demands answers to important questions — such as whether white or whole wheat bread toasts more quickly — a responsible scientist must set other work aside and carry out the pertinent experiments.

I was certainly not the first to study the chemistry of toast. That honor goes to Louis-Camille Maillard, who actually had a more ambitious project in mind: he wanted to unravel the secret of life. In 1912, Maillard developed an interest in how amino acids combined to form proteins in the body. He tried heating mixtures of amino acids, but without much luck. Then he decided to mix in some other chemicals that could reasonably be found in cells, such as fats and sugars, to see if these would somehow enhance the reaction. As soon as he added glucose, the most common sugar found in the body, the solution turned brown. This was due to a mixture of compounds, many of which contained nitrogen furnished by the amino acids. Since amino acids and sugars are present in the body, Maillard suggested that this reaction could have consequences for human health. But the greatest practical importance of his discovery, he thought, would pertain to the gingerbread and beer industries, because he was now convinced that the colors of these products were due to the "Maillard reaction."

As it turned out, not only did the high-temperature reaction between amino acids and sugars cause the gingerbread and beer to turn brown, but it also explained the brown hues of baked bread, coffee, soy sauce, and roasted meats. Even more important, the products of the Maillard reaction produced flavors; in fact, they produced a large variety of flavors because food contains a diversity of amino acids and sugars. By the middle of the 1900s, chemists at General Foods had investigated the Maillard reaction in such detail that they were able to produce specific flavors by manipulating the appropriate sugars and amino acids. When they added serine to hot glucose, they created the smell and taste of maple syrup. Cysteine produced the characteristic taste of roasted meat. Indeed, the secret to making a tasty stew is to brown floured meat (the flour is a sugar source) in hot oil before diluting the mixture with liquid. The Maillard reaction requires a temperature higher than the boiling point of water — if water is present, then the temperature cannot rise above its boiling point.

Artificial flavors had great commercial appeal, so chemists remained busy investigating numerous combinations of amino acids and sugars. One of these combinations, lysine and glucose, emitted the characteristic smell and flavor of toast — the very reaction I reproduced in my kitchen. The more lysine I added, the more the house smelled of toast. Now I was ready to tackle the problem I had set myself. My prediction was that whole wheat bread, which contains more lysine, a common amino acid in wheat gluten, would toast more quickly. Alas, how often do we see a beautiful theory ruined by one simple experiment! I observed no significant difference in the toasting rates. I perused the labels of the two breads. The white bread did indeed contain less protein, but it had ten percent more sugar. And there was another factor I hadn't considered: moisture content. If the whole wheat bread had a higher water content, its browning rate would be reduced. So I placed both breads in a vacuum chamber (this time in a real laboratory) and pumped out the water. The whole wheat bread did contain somewhat more than the white. Apparently, these factors conspired to prevent my whole wheat bread from demonstrating its championship toasting potential. I think that the extra sugar in the white bread compensated for its lower protein content.

Sugar not only contributes to the Maillard reaction, but it also caramelizes, thereby deepening the brown color. I confirmed this by toasting raisin bread — not just any raisin bread, but one made with "soluble raisin syrup," which boosted the loaf's sugar content to about six times that of white bread. I had hardly popped a slice into the toaster before the smoke alarm started blaring away. Raisin bread was clearly the toasting champ.

By now I'd had my fill of toast, but I was stuck with a pile of leftover experimental products. Wondering how to dispose of the stuff, I remembered reading that in the 1700s the English would steep toasted bread in water to make a "restorative" beverage. It cured all kinds of ailments. Or so they claimed. (Today we tend to worry about consuming too many burned foods, because they contain carcinogens.) I felt that after eating all that toast, a restorative was in order. Gathering up my remaining toast pieces, I tossed them into a glass of wine — just like they did in Shakespearean days to improve the wine's flavor; they would then raise their glasses and offer a "toast" to celebrate important moments. I can tell you that it tasted a lot better than either the natural toast water or my synthetic analogue. Its restorative properties remain to be tested, but maybe one day we'll be drinking a toast to toast wine.

GASSING GREEN BANANAS

If you have a headache, just take a banana and split it. Then tape half the peel to your forehead and the other half to the back of your neck. I learned about this high-tech therapy from none other than Ann Landers. One of the advice columnist's correspondents suggested that it was a practically foolproof method of getting rid of headaches. Doesn't sound foolproof to me. In fact, I think that a person who walked around with banana peel duct-taped to his head would look like a fool. But apparently that doesn't deter people who are convinced that the therapy works. And such people do exist. I know. I've spoken to one. A desperate one.

A lady who often resorted to the banana headache therapy called me one day in a panic. She had received an e-mail about the risk to Americans posed by Costa Rican bananas. These fruits, the letter asserted, harbored the strep A bacterium, which is responsible for necrotizing fasciitis, better known as flesh-eating disease. All are advised, the writer said, to stay away from bananas, unless they want to end up as a heap of festering flesh. Needless to say, this is total nonsense. Flesh-eating disease is not transmitted through food. This is not the only myth I've heard spun about the safety of bananas. A gentleman once informed me that bananas that had turned black were poisonous. Another insisted to me that this wasn't so, but that green bananas made you sick. Actually, the only thing these colors do is provide us with some interesting chemistry.

The banana, believe it or not, is the most popular fruit in North America. And it isn't even a fruit; botanically speaking, it's a herb. But never mind. Fruit or herb, bananas are picked green, because if they weren't they would never make their way into our mouths, or onto our foreheads. They would be over-ripe. The green color is chlorophyll, the plant pigment that is the key to photosynthesis. As a banana ripens, its need for photosynthesis decreases, and the membranes around the chloroplasts, the cells that produce chlorophyll, begin to break down. This allows the enzymes that metabolize chlorophyll to enter, and the green color gives way to the yellow of various anthocyanins. This cascade of events is triggered by the release of the plant hormone ethylene, a process that banana producers mimick by ripening their green produce with ethylene gas after transport. Indeed, you can make use of this bit of chemistry at home. Place green bananas in a bag with a yellow one, and the green will quickly turn yellow.

As a banana ripens, it goes through other changes as well. Its starch converts to sugar. A ripe banana has the equivalent of about five teaspoons of sugar, almost double that of a chocolate bar. To most people, a brown or black banana peel and the bruised fruit inside it are decidedly unappetizing. If a banana is bruised, it will quickly turn brown, but all the poor fruit is trying to do is protect itself. As far as the banana is concerned, a bruise is an attack on its integrity. It cannot tell the difference between an attack by fungi, insects, or humans — who may attempt to duct-tape its peel to their heads.

Any sort of injury causes chemicals called polyphenols to leak out of cells and come into contact with an enzyme known as polyphenoloxidase, or phenolase. The polyphenols are converted to quinones, which apparently have antifungal and insecticidal properties. Quinones can also react with each other to form giant molecules, or polymers, which are brown. This, in fact, is similar to the chemistry that occurs in our skin upon exposure to the sun. Damage to bananas can also result from cold temperatures — like those found in refrigerators or freezers. So much phenolic polymer forms that the banana skin turns black. It may look terrible, but it's still perfectly safe to eat.

In the early 1980s, quite a stir was created at McGill University by researchers examining the claims made by a healer named George Ille. The man had said that he could demonstrate his power by waving his hands over bananas and causing their color to change. He even had photographs to prove that the bananas subjected to his hovering hands ripened more quickly than a bunch of controls. I don't know how good a healer Ille was, but he could have made a fortune in the fruit business.

We should eat plenty of bananas, discolored or not. They are a great source of potassium, which can lower blood pressure. One study has even indicated that low levels of potassium in the blood may correlate with the risk of stroke. Bananas are effective against diarrhea as well. They absorb bile acids, which are linked with loose stools. You can even make use of the peel of a banana after you've finished eating the fruit. The peel contains amyl acetate, a good solvent. I've used banana peel to clean my shoes.

There's another interesting aspect to the banana peel. Surprisingly, perhaps, some of the polyphenols that are ultimately responsible for its dark discoloration have some effect upon the central nervous system. Serotonin, dopamine, and norepinephrine are all recognized neurotransmitters in the human nervous system. Banana peels do not contain much of these substances, but their presence perhaps explains why hippies in the 1960s tried to smoke banana peels to get high. But all those adventurous hippies got was indigestion. And maybe a headache.

So, we are back to that. Can there possibly be anything to the idea of treating headaches with banana peel? A chemistry teacher actually responded to the question in Ann Landers's column, suggesting that potassium absorbed through the skin could cure headaches if these were caused by hypertension. Makes about as much sense as the Costa Rican flesh-eating bananas. But I wouldn't rule out the possibility that tying anything to the head will help. Maybe binding the head collapses painfully distended blood vessels in the scalp, relieving a vascular headache. Put that in your pipe and smoke it.

SUMMERTIME BUCCANEERS

What would summer be like without sizzling steaks, charcoal-broiled hamburgers, or grilled hot dogs? Well, for one thing, it would probably be a little healthier. Let's find out why. The word *barbecue* derives from the ancient Caribbean tradition of cooking food over a fire on a scaffold made from green wood, a device that the Spanish named "barbacoa." They referred to the technique itself as "boucan." Shipwrecked sailors and runaway servants who found themselves on Caribbean islands picked up the method, and these people came to be called "boucaniers" — or, in English, "buccaneers." Many of us still turn into summertime buccaneers, brandishing our long forks and spatulas as we grill, roast, broil, and, of course, burn food to our stomach's content. But are we being a little too cavalier with our health?

As the saying goes, if you play with fire, occasionally you'll get burned. Especially if you try to rekindle smoldering charcoal with a squirt of lighter fluid. The fluid stream can catch fire and ignite the whole container, as well as its startled holder. There's more. Lighter fluid's volatile hydrocarbons can escape into the air, with about half the fluid evaporating before you ever strike a match. These vapors are unhealthy, and they contribute to smog formation. Self-igniting charcoal is not a useful alternative; it comes saturated with flammable hydrocarbons. The pollution effect is not trivial. During an average day in southern California, barbecues emit three to four tons of hydrocarbons, nitrogen oxides, and carbon monoxide — about the same daily emission rate as a petroleum refinery.

The best way to light a charcoal fire is with an electric starter or a newspaper tinder chimney. Real charcoal, made by heating wood in the absence of oxygen, is easier to light than briquets; it burns hotter, produces fewer noxious vapors, and leaves less residue. Briquets are made from crushed, charred wood scraps to which fillers such as starch or coal have been added. When we use these, we should only begin to cook after a uniform covering of gray ash has formed, indicating that the fillers have all burned away.

Gas barbecues burn much cleaner and hotter than charcoal ones. As long as it receives sufficient oxygen, propane will convert primarily to carbon dioxide and water vapor. The gas flame should be mostly blue, with some yellow at the tip. Too much yellow in the flame means that there is incomplete combustion. The yellow color is caused by the glowing pieces of soot that form when a partially blocked gas pipe prevents proper mixing of the propane with oxygen. Soot is an intricate network of carbon atoms with a very large absorbent surface area. It can absorb some of the unhealthy components of incomplete combustion and deposit them on the food. We must therefore thoroughly clean the cobwebs and soot from gas pipes at the beginning of each barbecuing season.

No matter what precautions we take, burning wood, coal, or meat will always produce some carcinogens. One of these compounds, benzopyrene — a polyaromatic hydrocarbon, or PAH — is so carcinogenic that researchers routinely employ it to induce cancer in the animals they use to evaluate new cancer treatments. There are several factors governing the amount of PAHs produced during barbecuing. These are the temperature at which we cook the food, the fuel we use, and the fat content of the food. Basically, the higher the temperature, the more carcinogens form.

Some outdoor chefs favor mesquite wood, because it imparts a unique flavor to the food. It does, however, exact a price. The major component of mesquite is lignin, which burns much hotter than the cellulose that makes up the bulk of most other woods. As a consequence, the smoke produced by mesquite contains more than twice as many polyaromatics as other wood smokes.

How can we minimize our exposure to these nasty polyaromatics? By precooking our meat in a microwave oven, we can minimize the time required on the grill. Barbecuing only low-fat foods, such as skinless chicken or fish, also helps. Barbecue sauce, due to its high sugar content, burns very easily, and we should therefore apply it to the food only near the end of the cooking process. The farther we position the food from the fire, the less likely it is to be contaminated with carcinogens. If we place the food too close to the fire, the outside will char quickly, but the inside will remain cool. The scientific explanation for this is that the outside cooks by heat radiation, and the inside cooks by conduction of heat via water. So our aim should be to place the food far enough from the heat source to permit the browning rate to match the conduction rate.

In most gas barbecues, the grill is at a fixed distance from the flame and cannot be adjusted. Lowering the flame, however, accomplishes the same effect as moving the food farther away. Skewering the meat on a metal rod — as we would do with a shish kebab — can also speed up the internal cooking. This is because the metal readily conducts heat into the food. Many backyard cooks barbecue with the lid down, as they've noticed that this makes the food more flavorful. Unfortunately, this is because the lid-down method involves more flaring up and more smoke, neither of which is desirable in terms of health. Finally, it is important that we clean the grill after each use to eliminate deposits of those nasty compounds. Oven cleaner works well — but let's face it, working with concentrated sodium hydroxide is not a particularly attractive prospect.

Given all of these concerns, why shouldn't we just forget the whole business? Because barbecued food tastes good. Specifically, it tastes good because high temperatures and smoke combine to produce very flavorful compounds. And, unfortunately, carcinogens. So does all of this mean that science is raining on our barbecues? No, but there may be a few sprinkles. The greatest risk involved is probably the fact that when we barbecue, we tend to eat too much fatty food. But the occasional barbecue — featuring low-fat foods grilled high above a source of low heat on a clean grill — is one of life's acceptable little pleasures. Just make sure that you don't put cooked food onto a plate that has held raw meat. This can result in food poisoning severe enough to make the most ardent backyard barbecuer leap for broccoli. Actually, eating broccoli along with barbecued food is a great idea. It contains sulphoraphane, one of the most potent anticarcinogens we've discovered so far. I wouldn't barbecue it, though.

AGITATE FOR ICE CREAM

Nancy Johnson of Philadelphia had a problem. She loved ice cream. But she found that making it was a struggle. She'd often spend up to an hour shaking the metal pot containing her mixture of cream and sugar before the stuff would freeze. And all that time she had to keep the pot immersed in a bath of ice and salt. There had to be a better way. So, in 1843, Nancy dreamed up the ice cream maker. She placed a metal can filled with ingredients in a wooden bucket and packed it with layers of ice and salt. Then she attached a hand crank to a brace positioned across the top of the bucket and ingeniously connected it to a paddle that would churn the mix as it froze. Thanks to Nancy, anyone could now make ice cream at home.

The concept of making ice cream is simple enough. Take some cream, add sugar and flavor, and freeze the mixture. Pure water freezes at zero degrees Celsius, but by dissolving any substance in water we lower its freezing point. So the ice cream mix, with all of its dissolved sugar, requires a temperature lower than zero to solidify. Now picture what happens if we place this mix in a container and then immerse it in a bucket packed with ice. The original temperature of the ice is well below zero (just check the temperature in your freezer), but the surfaces that are in contact with the air will quickly warm up to zero degrees and begin to melt. The water from the melted ice will also be at zero degrees, and this mixture of ice and water will remain at that temperature as long as any ice is present. But at zero degrees, the ice cream mix will not freeze. However, if we sprinkle salt on the ice, we create a different scenario. As before, the surface of the ice warms up and melts. The water dissolves the salt, and soon the pieces of ice are swimming in salt water. Since this liquid has a lower freezing point than pure water, the ice will lower its temperature until the new freezing point is reached. In other words, the ice cream container is now surrounded by salt water, which is at a temperature well below zero. The mix freezes.

But just freezing the mix won't give you ice cream. It will yield a dense, solid mass filled with ice crystals. Hardly mouth-watering stuff. If you want good taste, you must agitate. Shaking or mixing the ingredients during the freezing process is the key to making good ice cream. This accomplishes two things. First, it minimizes the size of the ice crystals that form; second, it blends air into the ice cream. The smaller the ice crystals, the smoother the ice cream. But it is the pockets of air blended into the product, known as the "overrun," that give it its prized foamy consistency. Nancy Johnson's hand-cranked device minimized crystal formation and incorporated air admirably. Indeed, ice cream manufacture today still uses the same principle.

Human ingenuity does come to the fore when ice cream makers are unavailable. During World War II, American airmen stationed in Britain and pining for ice cream discovered that the gunner's compartment in a bomber had just the right temperature and vibration level for making the sweet treat. They would put the ingredients into a can before a mission, stow it in the gunner's compartment, and then look forward to returning to base with a batch of freshly made ice cream.

The method of simultaneous mixing and freezing solves the main problems of ice cream manufacture, but it does introduce a complication. Cream essentially consists of tiny fat globules suspended in water. These globules do not coalesce, because each is surrounded by a protein membrane that attracts water, and the water keeps the globules apart. Stirring breaks the protein membrane, the fat particles come together, and the cream rises to the top. This effect may be desirable when we're making butter, but not when we're making ice cream. There is a simple solution: we can add an emulsifier to the mix. Emulsifiers are molecules that take the place of the protein membrane, since one end dissolves in fat, and the other in water. Lecithin, found in egg yolk, is an excellent example. That's why even the simplest ice cream recipe requires some egg yolk.

There is nothing like freshly made ice cream. Its smooth, airy consistency affords us a break from reality; it's a throwback to childhood and a less complex world. Storing ice cream, however, does present a problem — the dreaded heat shock. By taking the container out of the freezer, for example, we may cause the surface of the ice cream to melt. When it refreezes, it forms larger ice crystals, resulting in the crunchy texture that so terrifies ice cream lovers. The commercial solution? Add some wood pulp.

Now, don't get all worried — we're not talking about adding sawdust to ice cream. Microcrystalline cellulose is a highly purified wood derivative that sops up the water as ice cream melts and prevents it from refreezing into crystals. Cellulose is indigestible, and it comes out in the wash, so to speak. Guar gum, locust bean gum, or carrageenan, all from plant sources, can also be used for the same purpose. Although lecithin is a good emulsifier, there are others that are more commercially viable. Mono- and diglycerides or polysorbates disperse the fat globules very effectively.

For those of you yearning for homemade ice cream but unwilling to deal with the salt and ice, here's a solution. Find a chemist friend who can provide you with some liquid nitrogen and supervise your activity. Place the mix in a Styrofoam container, add liquid nitrogen, and stir. The mix freezes almost immediately and develops just the right foamy consistency as the nitrogen evaporates. With a little practice, you can outdo Nancy Johnson.

A final word of warning, though. Ice cream may be addictive. A study conducted at the U.S. Institute of Drug Abuse suggests that eating it stimulates the same receptors in the brain as certain drugs. If you run into this problem, you may want to sample one of the new flavors that commercial manufacturers are tinkering with. Garlic, spinach, pumpkin, or tuna ice cream is sure to curb your craving.

Man Cannot Live on Corn Alone

Italian cuisine is one of my favorites. Except for polenta. I have never developed a taste for that odd corn mush, which was once a dietary staple of poor Italians. When explorers returned home to Europe from North America with corn, it quickly became popular with landowners because of its abundant yield. These landowners often paid the farm workers they hired to grow the corn with a share of the crop, and corn became an important part of their diet.

By the late 1700s, however, it was becoming evident that the sharecroppers who subsisted on corn were an unhealthy bunch. One could easily recognize them by their crusty, reddened skin. "Pellagra," from the Italian for "rough skin," became a common term for the condition. Most people believed that it was caused by eating spoiled corn. Rough skin was not the only symptom the poor sharecroppers had to worry about. The disease was often characterized by a red tongue, a sore mouth, diarrhea, and dementia — before it killed its unfortunate victim. Pellagra came to be referred to as the disease of the "four Ds": dermatitis, diarrhea, dementia, and death.

By the early 1900s, pellagra had reached epidemic proportions in the southern U.S. It ravaged the poor, especially cotton pickers. Some sort of a communicable infection now seemed a more probable cause than contaminated corn. Rupert Blue, the U.S. Surgeon General, stepped in and assigned his top epidemiologist, Dr. Joseph Goldberger, to solve the mystery of pellagra. Many of the pellagra victims ended up in insane asylums, so these institutions seemed appropriate places to start the investigation. Goldberger soon realized that while many inmates had the symptoms of pellagra, no doctor, nurse, or attendant showed signs of the disease. He noted the same phenomenon in orphanages, where children often developed pellagra but staff members never did. This was inconceivable if pellagra were an infectious disease. So Goldberger began to ponder the lifestyle differences between the asylum and orphanage inhabitants and the attending staff of these institutions. He also began to speculate about differences in diet.

Goldberger observed some pretty dramatic differences. Both inmates and staff got plenty of food, but the variety was not the same. While the attendants dined on milk, butter, eggs, and meat, the pellagra sufferers had to subsist mostly on corn grits, corn mush, and syrup. Goldberger suspected that some sort of dietary deficiency might be triggering pellagra. But he uncovered one troublesome finding. In one orphanage he studied, most of the younger children showed symptoms of pellagra, but the older ones seemed to fare much better. This mystery was solved when Goldberger discovered that the resourceful older children were supplementing their diet with food that they snitched from the kitchen.

It was obvious to Goldberger what the next step in his investigation had to be. He must obtain government funding to add meat and dairy products to the diets of the orphans and the asylum inmates. He did so, and the results were miraculous. Almost all of the pellagra victims regained their health. But if he was to prove the dietary connection conclusively, Goldberger would have to conduct one more critical experiment. He would have to show that pellagra could be induced by a faulty diet. And where was he going to find volunteers for such a study? In prison. Convicts would do anything to get out of jail. So Goldberger approached the director of the Rankin Prison Farm in Mississippi and outlined his idea. The director agreed to cooperate. He would release any prisoner who volunteered to take part in Goldberger's study upon the study's completion.

The volunteers were soon lining up to lend Goldberger a hand, especially after the doctor explained the protocol. To the prisoners, it sounded like a cakewalk. For six months, they could eat to their heart's content, as long as they confined themselves to a menu of corn biscuits, corn mush, corn bread, collard greens, and coffee. Then they would be freed. After about five months, though, the fun went out of the experiment. The convicts began to suffer from stomachaches, red tongues, and skin lesions. Goldberger had proven his point. Unfortunately, he did not have a chance to cure his patients, since, true to his word, he'd had them released. The convicts quickly scattered, wanting no more of Goldberger's dietary schemes.

It would seem that the problem of pellagra was solved. But many scientists who had pet theories about contagion remained unconvinced. In a letter to his wife, a frustrated Goldberger described these colleagues as "blind, selfish, jealous, prejudiced asses." He would show them that pellagra was not a contagious disease! Dr. Goldberger organized a series of "filth parties," at which he swallowed and injected himself, his wife, and supportive colleagues with preparations made from the blood, sputum, urine, and feces of pellagra patients. Nobody came down with the disease. Goldberger had made his point by eating excrement.

Unfortunately, Dr. Goldberger did not live to see the day when the "pellagra-preventative factor" was finally identified. In 1937, scientists put the finger on niacin, one of the B vitamins. Corn, as it turns out, is a very poor source of niacin; when people — like Goldberger's inmates and orphans and convicts — eat little else, they develop pellagra, a deficiency disease. It's a shame that the convicts dispersed before the doctor could arrange to follow them up. It would have been interesting to see how they eventually fared, whether they suffered strokes or age-related macular degeneration, a leading cause of visual impairment. Why? Because recent studies demonstrate that lutein, a pigment abundant in corn, may be protective against both of these conditions.

Ultrasound measurements of the thickness of carotid arteries, a predisposing factor for stroke, reveal an inverse correlation with blood levels of lutein. Furthermore, lutein-fed mice that were genetically engineered to develop atherosclerosis developed lesions only half as large as those seen in mice on normal feed. Epidemiological studies have also shown that people who consume foods rich in lutein have a lower risk of macular degeneration. Apparently, lutein concentrates in the eye and protects it from the harmful effects of blue light. Sounds pretty good. It almost makes polenta sound appealing.

LESSONS FROM POPEYE

The most famous landmark in Crystal City, Texas, is a statue of Popeye the sailor man. He's squeezing his trademark can of spinach, ready to save Olive Oyl from the clutches of Bluto. Crystal City, you should know, is the spinach capital of the world. Its citizens erected the statue in 1937 to honor the character who single-handedly boosted spinach consumption and helped save an industry. But there may be more of interest in Crystal City than Popeye's statue. I think someone should look into the incidence of heart disease there. In Crystal City, spinach is a way of life — and, I suspect, a longer one. That's because spinach is an outstanding source of folic acid, a B vitamin that is increasingly being linked with a plethora of health benefits. Let me explain.

Our story starts in the hallowed halls of Harvard University, far from the spinach fields of Crystal City. It was here, in 1969, that Dr. Kilmer McCully became involved in the unusual case of a boy who died at the age of eight from a stroke. The boy had suffered from a rare condition in which a substance called homocysteine builds up in the blood. Homocysteine is a normal metabolite of methionine, a common amino acid found in virtually all dietary proteins. A healthy person's body quickly processes it, but it accumulates in those suffering from homocystinuria, like McCully's young patient. An autopsy clearly revealed the cause of death. The boy's arteries were like those of an old man. Could the damage have been caused by excess homocysteine, McCully wondered? To investigate this further, he needed to examine other children who were afflicted with the same condition.

It didn't take him long to reach a conclusion: children with high homocysteine levels sustain artery damage typical of that seen in older men. And then, to prove his point, McCully injected homocysteine into rabbits, causing artery damage. This was enough evidence to suggest a revolutionary idea: homocysteine is a risk factor for heart disease. McCully proposed that high levels of the substance cause damage quickly, while levels that are only slightly elevated take longer to wreak havoc. Excited by his findings, he submitted a paper to *The American Journal of Pathology*. But instead of getting famous, McCully got sacked.

Harvard denied him tenure, supposedly because of his unorthodox theory about heart disease. Members of the medical establishment had declared that cholesterol was the main culprit, and they could see no room for homocysteine in their scenario.

Eventually, however, Dr. McCully would be vindicated. And, somewhat fittingly, one of the first studies to show the validity of the homocysteine theory was carried out at the Harvard University School of Public Health. In 1992, researchers reported on an analysis of disease patterns in over fourteen thousand male physicians. Those subjects whose blood levels of homocysteine ranked in the top five percent had a heart attack risk that was three times greater than the risk calculated for subjects with the lowest levels. Numerous other studies have shown a similar relationship. A high homocysteine level (above twelve micromoles per liter) seems to be a clear, independent risk factor for heart disease.

Knowing about a risk factor is not much good unless we can do something about it. And in the case of homocysteine, we can. Let's take a moment to explore the relevant biochemistry. Homocysteine forms through the action of certain enzymes on methionine. Once it has formed, one of two things will happen. It will either be reconverted to methionine or metabolized to glutathione, a powerful antioxidant. Both of these pathways require the presence of B vitamins. The body needs folic acid and vitamin B_{12} to change homocysteine back to methionine, and it requires vitamin B_6 for the glutathione route. You are probably starting to get the picture. Inadequate levels of these B vitamins lead to increased levels of circulating homocysteine, which in turn causes arterial damage and heart disease.

The B vitamin doses we need to keep homocysteine in check are not extreme. About four hundred micrograms of folic acid, three micrograms of B_{12}, and three milligrams of B_6 should do the job. While we can certainly get these through diet, the fact is that many of us don't. Indeed, the average intake of folic acid in North America is about two hundred micrograms — far from adequate. This is where spinach comes in. It is an outstanding source of folic acid, particularly if we eat it raw. So, go for that spinach salad. And may I suggest dressing it with orange juice? Just one cup contains two hundred micrograms of folic acid. You can also throw in some green beans or cooked brown beans, also great sources of folate. You'll be helping your heart, and other parts of your anatomy as well. A recent study of 25,000 women showed that those who consumed the most folic acid were one-third less likely to develop precancerous polyps in their colon. And if that isn't motivation enough to seek out foods that are rich in folic acid, then consider that it may even lower the risk of Alzheimer's disease. Yup, you heard right.

Researchers at the University of Kentucky explored the Alzheimer's connection because they were aware of the extensive evidence showing that women who took folic acid supplements during pregnancy had babies with fewer neurological birth defects, such as spina bifida. Could folic acid affect the nervous system later in life, the researchers wondered? A group of nuns in Minnesota who had willed their bodies to scientific research provided the answer. Those who had ingested adequate amounts of folic acid throughout their lives were less likely to succumb to Alzheimer's disease. This finding was corroborated by researchers at Tufts University, who fed spinach to rats and found that it not only prevented but also reversed memory loss. It seems that homocysteine can damage nerve cells the same way it damages blood vessels.

What all of this comes down to is that Popeye was right. That's why I'm dismayed by his fading popularity among children. We could use his nutritional support. Especially when you consider that some researchers suggest we could prevent fifty thousand heart attacks a year in North America simply by increasing our folic acid intake. That spinach salad with orange dressing is looking mighty good.

Paprika's Peppery Past

I have to own up to a crime. I'm a thief. A paprika thief. And it's time I confessed and shed the burden of living with this misdeed. I committed this crime at the famed Gundel Restaurant in Budapest after finishing what was probably the best meal of my life. I'd had veal paprikash many times before, and I make a pretty good version of the dish myself, but the paprikash I was served at this temple of Hungarian cuisine was clearly superior. It had just the right degree of zest and hint of zing. And a stunning red color. There was doubt about it — I had to speak to the chef.

I had to do a little buttering up, but the immaculately dressed waiter agreed to introduce me to a sous-chef. (The head chef, it seems, does not speak to mere mortals.) I started by complimenting him on the exquisite dill pickles we were served, peeled no less, before getting down to the all-important paprika question. What variety did they use? He was tight-lipped about this, but implied that Gundel's took absolutely no risk with their paprika. No risk? That struck me as a curious statement. But I soon got the whole story.

Paprika, made from ground red peppers, is to Hungarian cuisine what tomatoes are to Italian, what curry is to Indian, or what soy is to Asian. Cooking without it is unimaginable. It is also unimaginable that any Hungarian would tamper with this national institution, but that is exactly what happened in 1994. A third of the paprika samples tested by the government that year were found to contain lead oxide. Officials launched the investigation when over forty people had to be hospitalized with lead poisoning after eating food flavored with paprika. The country went into a state of shock. Sales of paprika were halted. The government of the nation that already had the highest suicide rate in Europe braced itself for the worst. Cooks staggered about aimlessly in their home or restaurant kitchens, and rumor had it that a few had actually resorted to buying Spanish paprika.

The police unleashed a huge manhunt, which netted fifty-nine suspects, thirty-seven of whom were eventually charged with paprika adulteration. Their motive? Greed. Hungarians consume about a pound of paprika per person per year — or roughly ten million pounds. Lead oxide, a red pigment, looks a lot like paprika, and it's cheap. The ideal extender, if you don't care about the health of the consumer. Restaurateurs, of course, do care about their clients' health as well as their palates. The image of patrons stricken with lead poisoning is not good for business.

Since using imported paprika was unthinkable, many Hungarian cooks believed they had only one option: they must purchase whole dried peppers and grind the things themselves. Now I understood the sous-chef's comment about Gundel's paprika being "risk free." Ever since the crisis of 1994, all paprika has been "risk free" due to a national testing program instituted to ensure that such a scandal could not occur again.

The paprika-adulteration episode cut deeply into the Hungarian psyche. The spice is not only essential to the taste buds but also a source of great scientific pride. After all, it played a major role in the discovery of a factor present in certain foods that can prevent scurvy. Most people have heard of the exploits of the eighteenth-century British physician James Lind, who virtually eradicated scurvy in the Royal Navy by supplementing the sailors' diet with limes. Historians have attributed many a naval victory claimed by the "Limeys" to the fact that these sailors, unlike their enemies, were protected from scurvy. But the story of the actual isolation of the scurvy-preventative factor is less well known than Lind's feat. Not until the 1930s did scientists finally manage to identify that magical factor. And they didn't find it in limes — they found it in Hungarian paprika.

Dr. Albert Szent-Gyorgyi was a Hungarian physician who, around 1925, became interested in plant chemistry. He had noted a similarity between the darkening of damaged fruit and skin discoloration in patients suffering from Addison's disease, an adrenal gland disorder. Was there any common feature here, he wondered? Certain fruits, like oranges, did not turn brown, and their juice prevented others from discoloring. Szent-Gyorgyi isolated the substance that prevented the browning, but he had trouble determining its molecular structure and suggested the name "Godnose" for it. This did not go down well, so he changed it to "hexuronic acid." The plot thickened when Szent-Gyorgyi discovered tiny amounts in the adrenal glands of cattle, a discovery that would eventually turn out to have no importance. But it did stimulate the doctor to do more research — for which he needed large amounts of hexuronic acid.

As luck would have it, Szent-Gyorgyi was offered a university position in Szeged, which just happens to be the paprika capital of the world. To live in Szeged is to be surrounded by the sights and smells of paprika production. Szent-Gyorgyi couldn't help but wonder if paprika, like oranges and limes, might also contain his hexuronic acid. Did it ever. Within a short time, Szent-Gyorgyi had isolated a kilogram of the stuff and determined that it was identical to the antiscurvy factor found in citrus fruits. He rechristened it "ascorbic acid." Today we know it as vitamin C.

Recently, we have even learned that the major red pigment in paprika, capsanthin, is a potent antioxidant. What more could one ask for? We have a colorful spice that is good for us and tastes great. Of course, some paprikas taste better than others, which is why I just had to snitch that sample from the shaker that adorned every table at Gundel's. I may use it some-day if I have a dinner guest who is important enough.

BRING ON THE CRUCIFERS

A British food critic once suggested that, compared to boiled cabbage, "steamed coarse newsprint bought from bankrupt Finnish salvage dealers and heated over smoky oil stoves is an exquisite delicacy!" I have never tasted coarse newsprint, steamed or otherwise, but given the choice, I would go for the cabbage. And I think you should as well. Why? Because I think we could all do with some more indole-3-carbinol.

You may not realize it, but food is the most chemically com-plex substance we encounter in our daily lives. Cabbage, for example, contains hundreds of different compounds. Some are responsible for its color, some for its aroma, and others for its taste. There are also vitamins and minerals, as well as a host of compounds with unknown functions. Some of these com-pounds, however, are beginning to yield their secrets to science. And we may reap the benefits. Here's how.

The human body is a fantastic machine equipped with a range of defense mechanisms to protect itself against undesirable chemical intruders. A variety of enzymes can either convert these intruders into less harmful substances or link up with them and cause them to be eliminated through the urine. A cell's genetic machinery cranks out these protective enzymes when potentially dangerous foreign substances activate receptors on the cell's surface. Way back in the 1950s, researchers noted that cancer-causing substances triggered the release of protective enzymes; unfortunately, in many cases the enzymes were unable to eliminate the carcinogen completely. Clearly, though, some test animals fared better than others, apparently because they had more efficient enzyme-producing systems. There are human parallels here. Not every smoker develops lung cancer. Why not? Do the lucky ones produce more protective enzymes? And, if so, can we foster this trait?

Researchers uncovered a clue: they noted that rats exposed to one carcinogen were more resistant to the effects of a second carcinogen. These rats appeared to be protected by the enzymes their cells synthesized in response to the first attacker. Obviously, we cannot expose ourselves to a carcinogen to protect ourselves against other carcinogens. But what if we could identify substances that have a chemical similarity to cancer-causing agents but are themselves not dangerous? Might they not trick cells into generating protective enzymes? By the 1960s, we knew that this was a real possibility. We discovered that chemicals in cabbage, as well as in other cruciferous vegetables (so-called because of their cross-shaped leaves), like broccoli, cauliflower, and Brussels sprouts, could stimulate the production of protective enzymes. Soon, researchers were focusing on one specific compound that had aroused interest because of its potential in the fight against breast cancer — namely, indole-3-carbinol.

The connection here is through estrogen, the female hormone that has been linked with tumor promotion. The relationship between estrogen and breast cancer, admittedly, is not a simple one. Laboratory studies have shown that estrogen, like many chemicals in the body, undergoes a variety of reactions after it is produced. Its metabolism, as these reactions are collectively called, can take two alternative routes. One produces 16-hydroxyestrone, which seems to be the culprit in terms of stimulating the irregular multiplication of breast tissue. Alter-

natively, estrogen can be converted into 2-hydroxyestrone, a compound that is relatively inert. Both of these conversions are governed by specific enzymes, levels of which can be affected by various factors. This is where indole-3-carbinol comes in. It stimulates the protective enzymes that take estrogen down the safe path, meaning that there will be less exposure of breast tissue to the nasty 16-hydroxyestrone molecules.

That's pretty interesting stuff. It's also pretty abstract for most people. And it's not quite compelling enough to send them rushing to the kitchen to boil cabbage. But wait. Mice develop fewer mammary tumors when exposed to indole-3-carbinol. Rats exhibit less endometrial cancer. But things get even more interesting when we learn that researchers actually fed four-hundred-milligram capsules of indole-3-carbinol to women on a daily basis (roughly equivalent to the amount in half a head of cabbage) and determined that it really did affect the way that their bodies metabolized estrogen. Within two weeks, their levels of 2-hydroxyestrone — the good stuff — shot up. In fact, the levels rivaled those found in marathon runners, who have a lower incidence of breast cancer. So that's what happened to the pill poppers. But what about eating cabbage itself?

Thanks to some Israeli researchers, we have an answer to that question as well. Eighty women on a kibbutz agreed to follow a diet rich in cruciferous vegetables and submit their urine for analysis. The ratio of 2-hydroxyestrone to 16-hydroxyestrone in their urine increased, suggesting protection against breast cancer. It would be interesting to follow these women for a number of years (probably not a difficult task — cabbage is a notorious gas producer) to see whether their rates of breast cancer turn out to be lower. There is a good chance that they will be — at least if we judge by some interesting epidemiological evidence from Germany and Poland.

Breast cancer rates in the former East Germany were significantly lower than those in West Germany, but after unification, the disease pattern started to equalize. While East German and West German lifestyles diverged in many ways, researchers did note that cabbage consumption was much higher in East Germany. This becomes even more meaningful in light of recent research carried out at the University of Illinois. There researchers attempted to determine why Polish women who moved to the U.S. had a higher breast cancer rate than women in Poland. Cabbage is a staple of the Polish diet, but Polish Americans tend to eat less of it. Was this a factor, the researchers wondered? Their next step was to stimulate test tube colonies of human breast cancer cells with estrogen and then add cabbage extract. The cabbage-treated cells grew more slowly. And it was not a question of using unrealistic amounts of cabbage extract; doses were those achievable by eating the vegetable. Furthermore, the experiments suggested that the effect was due not only to indole-3-carbinol — other antiestrogenic compounds also seemed to be present in the cabbage juice.

Now, I guess, we're ready to head for the kitchen. Especially when we consider that cabbage is also high in vitamin K, which is receiving more and more attention for its role in strengthening bone. Researchers conducting the Nurses Health Study, which has been following over seventy thousand women for more than ten years, found that those who consumed moderate to high amounts of vitamin K from vegetable sources had a thirty-percent-lower risk of hip fractures. Still need more convincing? Then think about the fact that epidemiological studies show that people who claim to eat cabbage regularly have a lower risk of colon cancer.

Ah, now I can smell the cabbage cooking. But do it right: don't boil it in water. That's how you release the smelly sulfur compounds. As a general rule, the more you cook cabbage, the worse it smells. So, just stir-fry shredded cabbage in a little olive oil until it turns brown and then cook it in its own steam for a few minutes. Add a little salt, a grind of pepper, and a touch of sugar. Then toss it with some freshly boiled thin noodles. Couldn't ask for anything better. Except perhaps the cabbage strudel my mother used to make. It was delicious and well worth the intestinal turmoil it sometimes caused. Unfortunately, I don't know how to make it, so I'm settling for something I can make. Cabbage pizza. Don't laugh. Try it. I'll guarantee that it's a delicacy — and not only when it's compared with steamed Finnish cardboard.

BEER SCIENCE IS STILL BREWING

Here's a question to ponder. When did civilization begin? When was it that we reached a sufficiently high level of cultural and technological development to call ourselves civilized? Looking at the world today, some would argue that we still have a way to go, but others point to beer as the first product of civilization. And they may be right.

The Sumerians, living five thousand years ago in what is now Iran, are commonly believed to be the world's first civilization. And they were into beer. In a big way. They brewed at least sixteen varieties, all from barley, just like today. We know this because a piece of Sumerian pottery, unearthed in 1992 and currently housed in the Royal Ontario Museum, harbored a yellowish residue that turned out to be calcium oxalate, a signature of beer brewed from barley.

The ancient Egyptians also raised barley and developed pure yeasts to make beer. They even learned to seal beer in jars to prevent a secondary fermentation. Beer was not only drunk for its alcoholic effect, it was also consumed for its supposed medicinal properties. The ancients treated scorpion bites with beer, albeit ineffectively, and they ate onions steeped in beer as a "remedy against death."

In the Middle Ages, beer was an extremely popular beverage. Fermentation yields an acidic brew that is not conducive to the growth of bacteria, so beer was certainly safer than the dirty, untreated water that most people were forced to drink back then. But even beer drinkers had a concern. As a drinker hoisted a mug of the brew, his view would be temporarily obscured, and any pickpocket in the vicinity would seize the opportunity to ply his trade. Some believe that this gave rise to the glass-bottom mug and to the expression "Here's looking at you!" Many were also concerned about the adulteration of beer. As early as the eleventh century, Edward the Confessor employed ale testers to check up on the brewers. These testers would spill some ale on a wooden seat and then sit on the puddle wearing leather breeches. If they had a hard time getting up, they knew the ale had been sugared. A definite no-no.

Indeed, the world's oldest consumer-protection law was formulated to ensure the purity of beer. In 1516, Count William IV of Bavaria issued the Purity Law, which allowed brewers to use only barley, hops, and water in making their beer. This law is still in force in Germany. And the Germans pride themselves on the quality of their beer. In fact, in 1870, Louis Pasteur transformed brewing from an art to a science when he resolved to make French beer as good as German. He discovered that the spoilage of beer was due to the presence of microorganisms foreign to the nature of the true beer yeast.

Unwanted microbes are not the only enemy of the brewer's art. The scent of 3-methyl-2-butene-1-thiol can be most disturbing. Unless you're a skunk. If you are, then the stuff is a valuable commodity, because it effectively wards off predators. But people certainly don't want it in their beer. How does it get in there? Skunks aren't to blame — light is. It may be hard to believe that light can alter the flavor of beer, but it most assuredly can. Hops — or, more specifically, the blossoms of the female hop plant — are the secret to the taste of beer. They were originally added to compensate for the sweet taste of malt, but they turned out to have value in the control of undesirable microbes. There are many compounds present in hops, including some that have estrogen-like effects and may cause breast growth in those who drink beer to excess. But one specific compound, iso-humulone, seems most important when it comes to solving the problem of "light-struck flavor." Light cleaves the molecule to produce an active fragment that then reacts with some sulfur compounds found in beer to form the offensive 3-methyl-2-butene-1-thiol.

We should not be surprised that molecules are affected by light. After all, we know all about the damage that sunlight can do to molecules in our skin. It can do the same to iso-humulone.

Is there a solution to this problem? Sure there is. We can block the rays that damage beer the same way we block the rays that damage our skin. Of course, applying sunscreen to beer bottles is not an option. But building protection directly into those bottles is. And that is why beer is sold in brown bottles. The brown pigment in the glass filters out the wavelengths of light that cause the skunky smell.

So why do some manufacturers bottle their product in clear glass bottles? Chemical ingenuity has made this nuance possible. Through the process of hydrogenation — much like the one we use for margarine — we can alter slightly the molecular structure of iso-humulone, making it stable in light. We can now admire the golden color of our favorite beer without even opening the bottle. The true beer lover, however, won't hesitate to crack open a cold one and gulp down a healthy dose. And it may be healthy, indeed, although much depends on the dose.

A study published in *The British Medical Journal* examined the beer-drinking habits of a group of people who had suffered heart attacks and the beer-drinking habits of a group randomly selected from the Czech population. The Czech Republic is especially appropriate for such a study, because it is a country where beer is the beverage of choice. Perhaps surprisingly, in both groups the lowest risk of heart attack was found among the men who drank nine to twenty pints a week. Their risk was a third of that seen in the men who never drank beer. But if they drank more, they lost that protection and developed problems. Dark beer seems to be especially protective. Researchers discovered that it even reduces the potential harm caused by the notorious heterocyclic aromatic amines (HAAS) that form when food is heated to a high temperature. Serving dark beer at a barbecue is a good idea. Maybe Benjamin Franklin was on to something when he said, "Beer is proof that God loves us and wants us to be happy."

The Scoop on Booze

The police officers could hardly believe their eyes. The eighteen-year-old driver they had just pulled over sat there speechless, a wad of white fabric sticking out of his mouth. He had ripped the crotch out of his underwear and stuffed it into his mouth in an apparent attempt to fool the Breathalyzer. A memory of someone saying that cotton is highly absorbent must have stirred in his confused mind, prompting this bizarre reaction. But the Breathalyzer was not fooled. Neither was it fooled by the teenager who started sucking furiously on pennies after the highway patrol stopped him. He must have remembered a bit of the chemistry he had learned in school — the bit about alcohol being oxidized to acetaldehyde by the action of copper. He figured he'd be in the clear, since the Breathalyzer tests for alcohol, and not acetaldehyde. Unfortunately, the young genius didn't remember accurately. Ethanol, the alcohol in beverages, can indeed be converted to acetaldehyde by copper, but only when the copper is red hot.

Then there are those who try to outsmart the police by insisting that they've just used mouthwash. But this doesn't wash either. Sure, mouthwashes contain alcohol, and a false Breathalyzer reading is possible, but only if the test subject rinses out his mouth immediately before giving a breath sample. The police must follow certain guidelines: they have to observe a suspect for several minutes before administering a Breathalyzer test, and alcohol from mouthwash dissipates within a couple of minutes.

Is it surprising that people resort to such curious acts when they've overindulged? Not really. After all, alcohol certainly affects the brain. And the rest of the body as well. The chemistry involved is absolutely fascinating. Before alcohol can affect the brain, it has to get there. Most of the alcohol we consume is absorbed into the bloodstream from the stomach and the small intestine. But not all of the alcohol makes it through. Some is metabolized in the mucosa that lines the stomach and intestine. Here, enzymes convert ethanol first to acetaldehyde and then to acetic acid, neither of which is inebriating. In men, about thirty percent of a dose of alcohol meets its metabolic end in this fashion, but there is a definite gender bias here. The female stomach and intestinal lining is only about half as efficient at breaking down ethanol, so more makes it into the circulation. This explains why women may become tipsy more easily.

Once the alcohol is in the bloodstream, it passes through the liver. The liver is the body's main detoxicating organ, and it detects alcohol as a potential troublemaker. First, an enzyme called alcohol dehydrogenase snips a couple of hydrogen atoms out of the ethanol molecule, converting it to acetaldehyde. Then aldehyde dehydrogenase transforms this intermediate into acetic acid, which is either excreted or used by the body as a source of energy as it is broken down into carbon dioxide and water. A gram of ethanol can provide about seven calories in this fashion. If a person's intake of alcohol is sufficiently high, the liver's detoxicating system becomes overburdened, and some of the alcohol slips through unmetabolized. It can then wreak havoc in the brain.

Ethanol does this by interfering with neurotransmitters, the chemicals that brain cells use to communicate with one another. At low alcohol levels, receptors for glutamate are activated, leading to stimulation and a loss of inhibition. This is the "social lubricant" effect of alcohol. But as the concentration of alcohol rises, glutamate receptors actually become less responsive; the drinker begins to slur his or her words, and "cocktail party amnesia" sets in. Other neurotransmitter systems are also affected. Gamma aminobutanoic acid (GABA) is known as an inhibitory neurotransmitter because it prevents nerve cells from firing excessively. Alcohol stimulates GABA activity, which eventually causes sedation and relaxation. And that is only part of a very complex picture.

Eventually, the effects wear off as the alcohol is excreted or metabolized as it passes through the liver again. But, as this is happening, the drinker must contend with nausea, headaches, and a flushed face. The culprit here is acetaldehyde, some of which escapes from the liver before being converted to acetic acid. Not everyone experiences these symptoms to the same degree. Many people of Asian origin are severely affected by facial flushing, because nature has dealt them a very slow-acting version of aldehyde dehydrogenase, the enzyme that normally degrades acetaldehyde. Indeed, the same phenomenon lies behind a prescription drug known as disulfiram (Antabuse), which physicians give to alcoholic patients. The drug inactivates aldehyde dehydrogenase, forcing acetaldehyde into the circulation. This should make the drinker so sick that he gives up the booze. Unfortunately, he usually gives up the drug instead.

Some of the effects of acetaldehyde can linger till the morning after and contribute to a hangover. Interestingly, scientific researchers have not investigated the hangover business as extensively as one would expect. That's because solving this problem would trigger a whole new problem. Some are concerned that if the hangover is eliminated, then people will drink more. Still, we do know that there is more to the hangover than just the remnants of acetaldehyde. The metabolism of alcohol in the liver produces some free-radical debris, which is usually taken care of by glutathione, one of the body's most important antioxidants. When the system is overwhelmed, these free radicals can contribute to the hangover. That is why researchers have had some success in treating hangovers with supplements of N-acetylcysteine (NAC), which serves as a source of cysteine, the critical compound the body needs to generate more glutathione.

Eggs also contain cysteine, and that may explain why traditionally people have used them as a hangover treatment.

The hangover is actually multifactorial. Dehydration plays
an important role, as does hypoglycemia caused by the alcohol-
mediated loss of sugar in the urine. But, in all likelihood, the
greatest contributor to the hangover is methanol. This alcohol
is found in small concentrations in many beverages; it's a by-
product of fermentation. Methanol is metabolized by the same
enzymes as ethanol, but the products this time are formaldehyde and formic acid, which produce the hangover symptoms.
Why does this happen only the morning after? Because the
enzymes prefer to work on ethanol instead of methanol. Only
when all the ethanol has been metabolized do they switch to
methanol. This then explains the "hair of the dog" hangover
remedy. A drink in the morning supplies ethanol for the
enzymes to act upon so they'll leave the methanol alone. As the
enzymes busily metabolize the ethanol, methanol is excreted in
the urine without being converted to formic acid. A Bloody
Mary may be the best choice here, because vodka contains very
little methanol. Confirmation of the critical role methanol
plays in hangovers comes from a study showing that treatment
with 4-methylpyrazole, a drug that blocks the breakdown of
methanol, can eliminate the symptoms.

I must admit to feeling a little queasy talking about hangover cures. Alcohol can be an extremely destructive beverage.
It is probably more damaging to society than all illicit drugs
combined. Cirrhosis of the liver, strokes, breast cancer, oral
cancers, domestic violence, and sexual assault have all been
linked to alcohol abuse. In North America, there is an alcohol-
related car accident every thirty seconds. And, as if that wasn't
frightening enough, excessive alcohol can shrink the genitals
and have feminizing effects on men. The male drinker produces
less testosterone, so his sex drive flags. But, for those who want

to look on the bright side, less testosterone means less likelihood of baldness.

Henny Youngman, whom some would call a comedian, once remarked that when he read about the evils of drinking he gave up reading. I hope you won't do the same. There is nothing funny about being drunk. Drunks destroy their own lives and kill others. What can we do? Well, University of Georgia researchers have found that blood alcohol can be reduced significantly by inserting a tube into the rectum and piping in alcohol dehydrogenase and oxygen. Sounds good to me.

Hard Lessons about Soft Drinks

Robert Southey was poet laureate of Britain for thirty years. He used his position and talent to attack the social structure of the times, and he was particularly critical of the idea of slavery. Most of us would be hard pressed to quote any of his works, but an expression that Southey coined has made it into our everyday vocabulary. In 1812, someone served the poet a new-fangled beverage made with soda water and ginger beer. The bottle was corked, and it made a popping sound when opened. In a letter to a friend, Southey described the fun he'd had with "soda pop," a beverage that we have been having fun with ever since. Let's face it, there are some things in life that we do purely for the fun of it. And drinking soda pop, or soft drinks, is one of them. But there may be a price to pay for the fun.

Soda pop was not always marketed as a refreshing beverage. Its early manufacturers actually touted it as a medicine. In the middle 1800s, North American doctors frequently diagnosed their patients with a condition they labeled "neurasthenia." It was a catch-all term that described occasional fatigue, insomnia, depression, and achy muscles — in other words, the symptoms of life. John Pemberton, an Atlanta pharmacist, thought he had a way to treat neurasthenia: French Wine Cola, a solution made with extracts of the coca leaf, the African kola nut (a source of caffeine), damiana root (a supposed aphrodisiac), and wine. The coca leaf is the source of cocaine, a substance that Pemberton was very familiar with. He had been wounded in the Civil War and was left with a permanent disabling pain that had caused him to become addicted to morphine. At the time, people believed that cocaine was an antidote for morphine addiction, and the substance was perfectly legal. Pemberton's acquaintance with the coca leaf prompted him to try it as an ingredient in his concoction. The amount that he used, however, was so small that the dose of cocaine in French Wine Cola was trivial.

When the temperance movement came to Atlanta and alcohol was banned, Pemberton had to reformulate his product. Instead of wine, he used extracts of orange, lemon, nutmeg, cinnamon, coriander, neroli, caramel, and vanilla blended with sugar and lime juice. All he needed now was a captivating name. His accountant suggested Coca-Cola, and it was destined to become the most famous trademark in history.

At first, Pemberton sold his Coca-Cola as a thick syrup that had to be mixed with water. He promoted it as a new elixir that would treat melancholy, hysteria, rheumatism, and obesity. People sang its praises, and some physicians even complained that the elixir was stealing their patients away from them. Then a pivotal moment arrived. A gentleman suffering from a terrible headache walked into a pharmacy and asked the pharmacist to mix him up a dose of Coca-Cola as quickly as possible. Since the patient seemed so distraught, the pharmacist thought he'd save a little time, and instead of walking to the other side of the counter where the water tap was located, he mixed the syrup with soda water, which he had on hand. Pharmacies in those days dispensed soda water because of its reputed health benefits.

This became possible after Jacob Schweppe designed the necessary equipment to carbonate water on a large scale, capitalizing on Joseph Priestley's 1772 discovery of infusing water with carbon dioxide. Not only did the Coca-Cola/soda-water concoction relieve the man's headache, but it also relieved his thirst. Based on this experience, the pharmacist began to mix Coca-Cola syrup routinely with soda water, and the world's most popular soft drink was born. Pemberton soon realized the potential of this mix, and Coca-Cola transcended its status as a medicinal tonic to become a "delicious, refreshing, exhilarating, invigorating" beverage that even the healthy could enjoy. Contrary to the popular myth, the exhilaration the drink provided was not a result of the cocaine — it was an effect of the caffeine-containing kola nut.

Coca-Cola was not the only beverage to tackle health problems. Other soft drinks ("soft" because they contained no alcohol) also got in on the game. In Waco, Texas, pharmacist Charles Alderton came up with Dr. Pepper, which he claimed could aid digestion and restore vim, vigor, and vitality. Charles Hires began to market a root beer that was sure to "purify the blood and make rosy cheeks." Later, Lithiated Lemon-Lime Soda was introduced with the slogan "Takes the 'ouch' out of 'grouch.'" The drink was a huge success, perhaps because it contained lithium, which has mood-altering properties. Eventually, the name was changed to 7-UP — presumably a combination of "seven ounces" and "bottoms up!"

But not all soft drinks were sold in bottles. Up until the 1950s, pharmacies had soda fountains. It was at the soda fountain that the soda jerk (so called because of the jerking motion he made when dispensing the carbonated water from its container) mixed up a variety of fizzy beverages. He would spoon out flavored syrup and add just the right amount of soda water to delight young and old alike.

Today, we guzzle roughly sixteen ounces of pop per person per day in North America — a staggering amount. Consumption has doubled since 1975. Teenagers are the heaviest consumers, with boys averaging close to three cans a day. Some researchers suggest that these heavy users are also becoming just plain heavy. Obesity is a growing problem in North America, a problem to which sugared soft drinks may contribute. One can has about seven teaspoons of sugar, which accounts for the drink's 120 or so calories. (Fruit juices have about the same amount of sugar, but they also provide a variety of vitamins, minerals, and antioxidants.) And serving sizes are increasing: many outlets — fast-food restaurants and movie theaters — now sell soda pop in virtual buckets. Furthermore, research has demonstrated that the body does not handle these liquid calories the same way it handles calories from other foods.

In one fascinating study, researchers asked subjects to consume 450 calories' worth of jellybeans daily for four weeks, and then 450 calories' worth of soda pop for the same period. During the jellybean feast, they ate about 450 calories less of other foods, but they didn't during the pop fiesta — they ended up increasing their total calorie intake by 450 calories. Scientists have also found that people who consume a sugar-sweetened drink with a meal eat more than they would if they chose a calorie-free beverage instead. And soft drinks tend to squeeze milk out of the diet, resulting in reduced calcium intake and a greater risk of osteoporosis. A University of Saskatchewan study showed that teenaged girls who drank soft drinks instead of milk already had a reduced bone-mineral content. This was not due, as some allege, to phosphates in colas robbing their bones of calcium, but it was clearly related to soft drinks displacing more nutritious drinks from their diets.

We obviously have legitimate reasons to limit our soft drink consumption, but we should be wary of the numerous nonsensical reasons that are floating around. Polyethylene glycol, used to disperse insoluble flavor components, is not antifreeze. The body does not absorb it, and it is harmless. Soft drinks do not acidify the blood and allow cancers to grow. Neither are these beverages as damaging to teeth as candies and sweetened desserts. A tooth will not dissolve in Coke within twenty-four hours, and a T-bone steak will not dissolve in Coke within forty-eight hours. The caramel coloring in colas is not a carcinogen. Phosphoric acid does not promote indigestion by fighting with hydrochloric acid in the stomach. Cold drinks do not reduce the effectiveness of digestive enzymes, causing food to be fermented instead of digested. Aluminum contamination from canned drinks does not cause Alzheimer's disease.

If you are looking for a real cola horror story, here's one. An usher at a movie theater popped out before the film started to buy an ice-cold cola from a machine. As he started to drink, he felt a round, solid object bob against his lips. He had a horrible thought: the foreign object could be a severed finger or a small animal. He immediately stopped drinking. Although the usher had to get back to work, he kept the can for later investigation. After the movie, he poured out the can's contents, finding nothing. But when he purchased another can from the same machine, the mystery was solved. That can contained a similar solid object: a finger of ice.

And here's yet another scary account. Deglutition syncope is a rare condition that can be caused by drinking cold, carbonated beverages; the drinker faints after guzzling a soft drink too rapidly. In some way, the cold drink stimulates the esophagus to put out a signal that affects the heart, causing it to beat more slowly. Blood pressure drops, dizziness and confusion set in, and, in extreme cases, the subject faints. Yet soft drinks can sometimes have a beneficial effect on the esophagus. Some unfortunate people suffer from a narrowed esophagus, and their food gets stuck on the way down. Often, a physician has to insert a scope into the patient's esophagus to clear the blockage. Occasionally, however, the patient can solve the problem by simply drinking a fizzy beverage. The drink penetrates the stuck food, and the escaping carbon dioxide then dislodges it. A good burp, and the food slides right down the pipe.

And Coke can do other useful things. Phosphate from phosphoric acid is a great rust remover, forming a soluble complex with iron. Coke can therefore be used to loosen rusty bolts. You can remove rust spots on a chrome bumper with aluminum foil dipped in Coke. In a pinch, you can use it to clean a toilet bowl as well. A ham wrapped in foil and treated with Coke will produce a delicious gravy, and, believe it or not, there is even a recipe for Coca-Cola cake. It's made with flour, sugar, butter, cocoa, buttermilk, eggs, baking soda, vanilla, marshmallows, and Coke. Now, if that doesn't take the cake, I don't know what does.

The modern soft drink industry is huge, producing about 110 billion liters of product every year. The basic ingredient is, of course, water, but not any old water. The water that goes into soft drinks is purer than the stuff that comes out of the tap. First, the manufacturer adds a flocculating agent, like aluminum or iron sulfate. This forms a gel-like material that settles to the bottom, taking impurities down with it. Chlorination follows to eliminate microbes, and then the manufacturer mixes in lime (calcium hydroxide) and soda ash (sodium carbonate) to precipitate out calcium and magnesium. We refer to this as removing the "hardness" minerals. If the drink maker does not reduce the amount of these alkaline minerals in the water, then they will interfere with the taste of the product by neutralizing the acids that are part and parcel of that taste. The water then passes through an activated carbon filter to remove residual chlorine and other off-flavors, along with any tiny particulates that can act as nucleation sites for carbon dioxide pockets and cause "gushing."

After water, the next most important ingredient is sugar. Traditionally, manufacturers used cane or beet sugar (sucrose), but today they are more likely to employ sweeteners derived from corn. Cornstarch is a giant molecule composed of individual glucose units strung together. Enzymes from easily cultured molds can turn cornstarch into corn syrup by breaking down the long glucose chains. The syrup contains glucose, maltose (two glucose units joined together), and molecules composed of a varying number of glucose units. Both glucose and maltose are sweet, and hence manufacturers can determine the sweetness of the syrup by controlling the extent of starch breakdown. But there is another clever method they can adopt to increase the sweetness of the syrup, allowing them to use less of it. Using enzymes from various species of the bacterial genus *Streptomyce*, which is sweeter than sucrose, they can convert glucose to fructose. So-called high fructose corn syrups are now frequent replacements for sucrose in soft drinks because they are cheaper. Also, manufacturers find the liquid form of the sweetener more convenient to work with.

Corn sweetener has its pros and cons. Sucrose, over time, breaks down to form glucose and fructose, altering the taste of the beverage. This does not happen with high-fructose corn syrup. But if we replace sugar with high-fructose corn syrup, we will end up consuming about two and a half times as much corn syrup as we did in 1980. Fructose metabolism has a higher requirement for chromium and copper than the metabolism of other sugars, and this can conceivably lead to a drop in chromium and copper blood levels. Rats fed a high-fructose/low-copper diet routinely develop higher cholesterol, triglycerides, and blood sugar.

While water and sugar make up the bulk of a soft drink, its essence comes from the flavor components. Chemists have identified some six thousand compounds with specific flavors, but the majority of flavor blends derive from only eight hundred or so compounds. Manufacturers go to great lengths to try to keep their formulations secret — they are determined to prevent copying. You might think that in this age of supersophisticated chemical analysis, one could easily uncover such secrets. But you would be wrong. Nobody has been able to duplicate the exact flavor of Coca-Cola, the famous formula 7X. The basic composition is well known; in fact, Pemberton's original notebooks have been found. What we don't know is the exact ratio of the orange, lemon, nutmeg, cinnamon, coriander, and neroli oils, and the order in which these are mixed with the vanilla extract, lime juice, and phosphoric acid. Chemical analysis can reveal the compounds that are responsible for the flavor, but it cannot tell us how the blend was arrived at. Imitators have come close to 7X, but so far, no cigar.

The complexity of the situation is exacerbated by the use of various additives. We can divide flavors into two categories: water soluble and water dispersible. Compounds such as vanillin and cinnamic aldehyde mix easily with water, but most fruit flavors do not. Manufacturers employ various polysaccharides — such as carrageenan, alginates, or tree gums — to disperse these flavors as miniscule droplets through a beverage. Brominated vegetable oils are a somewhat more controversial class of additives. Incorporating bromine into the molecular structure of a vegetable oil increases its density, so that a flavoring oil dissolved in it will not rise and separate from the water. Drink makers commonly use it in fruit-flavored beverages; it has the added effect of producing cloudiness, making the soft drink resemble fruit juice. The results of animal tests have raised some eyebrows, because when brominated vegetable oil is deposited in tissues such as the heart, there can be adverse effects. The amounts that researchers used in these studies, however, were far greater than the amounts humans are exposed to. Sucrose acetate isobutyrate mixed with vegetable oil is an alternative to brominated vegetable oils, and many manufacturers favor it. It is excellent in creating stable, cloudy beverages, and it effectively prevents flavoring oils from rising to the top and forming an unsightly ring at the neck of the bottle.

Some of the flavor compounds in soft drinks are susceptible to oxidation. Manufacturers counter this effect by adding antioxidants such as ascorbic acid (vitamin C). Preservatives such as sodium or potassium benzoate ensure that microbes do not contaminate the drink, and approved artificial colors or caramel will increase eye appeal. In rare cases, some of these colors — particularly yellow #5 and yellow #6 — can cause allergic reactions.

Some people object to these additives, but there are far better reasons for not gorging on soft drinks than the fact that they may contain brominated vegetable oils or food dyes. We don't need the extra sugar or caffeine. We do need the calcium in milk and the antioxidants in fruit juices, and soft drinks often displace these beverages from the diet. But we do not have to shun soft drinks totally. After all, they do add a little much-needed fun to our lives. Let's just try not to have too much fun.

WHEN DNA COME OUT TO PLAY

"Wow! Look at my DNA!" the exuberant little boy cried as he pulled the threadlike strands out of the test tube. Soon, other excited voices chimed in as about two dozen children and a sprinkling of adults began to play with their own genetic material. We were all seated around tables in a laboratory at the

American Museum of Natural History in New York City, having been attracted by signs pointing towards "The Gene Scene." Our experiment started with everyone swirling salt water in their mouths for thirty seconds or so to collect some of the cells that our cheeks continuously slough off. Our guide then asked us to spit the solution into a little cup (cries of "Yuck!" filled the room) and then transfer it to a test tube containing some detergent.

A couple of minutes of gentle shaking allowed the detergent to break down the cell membranes and liberate the DNA molecules, which formed a precipitate when a little alcohol was added. Next, we dipped a stirring rod into our test tubes and pulled out long filaments of DNA. As the session drew to a close, the guide asked the children present what they had learned. They provided some pretty good answers, but the one that really stuck in my mind came from the little boy who had cried out so enthusiastically when he first glimpsed his DNA. He said he'd learned that when he grew up he wanted to study biotechnology and become a genetic engineer. Quite a refreshing comment, given that so many people these days look warily on this area of science. We can generally attribute this reaction to confusion over what biotechnology is really about.

Simply put, biotechnology is the provision of useful products and services from biological processes. It does not necessarily involve scientists in white lab coats hovering over petri dishes. In fact, biotechnology goes back thousands of years. It probably began the first time someone used yeast to convert sugars and starches to alcohol. Yeast is a little living machine that takes in food and produces excrement. But don't pooh-pooh that excrement. Many humans like it. It's called alcohol. Molds are also neat little machines that produce a variety of by-products. When the ancient Egyptians applied moldy bread to wounds as a poultice, they were exploiting biotechnology. The mold probably churned out penicillin — which, of course, the ancients did not recognize as such — and it helped heal the wound.

No one was able to elucidate how these microbes convert raw materials into finished products until relatively recent times. The pivotal moment came in 1953, when Francis Crick and James Watson unraveled the molecular structure of DNA, the molecule that controls the inner workings of the living cell. The instructions for everything that a cell does are encoded in genes, which are specific fragments of DNA. Basically, genes tell the cell what proteins to produce. Proteins are needed as structural material and as enzymes, the catalysts that control all reactions in a cell. Once we clearly understood DNA's role, it became obvious that if we could modify its structure, then we could alter the proteins it produced. By the 1970s, such manipulation — known as genetic engineering — had become a possibility. Scientists could transfer genes from one organism to another or even build them from fundamental components using the "Gene Machine" (invented by former McGill University chemistry professor Kelvin Ogilvie).

We are now beginning to see some of the practical results of this genetic tinkering. For example, cheese makers require an enzyme called chymosin to separate curds from whey. The traditional source is the stomach lining of calves, but we have now isolated the fragment of DNA, the gene, that tells the cell to produce this enzyme. We can incorporate it into the DNA of a yeast, which will then dutifully crank out chymosin. This has made cheese production more efficient, and it has also allowed for the manufacture of cheese that has no meat components — a factor that is desirable to those who conform to certain religious dietary restrictions, such as kashruth. Much more dramatic is the potential for treating people who suffer from specific immune system deficiencies due to a malfunctioning gene. Already, in one case, doctors have extracted a patient's bone marrow, replaced the malfunctioning gene, and infused the marrow back into the bone. The patient now produces cells with normal genes.

These days, we use bacteria to which the human insulin gene has been transferred to crank out insulin for diabetics. Scientists have also engineered bacteria to produce TPA (tissue plasminogen activator), which has saved countless lives. Physicians commonly administer it to patients who have suffered heart attacks to dissolve blood clots. Unfortunately, bacterial fermentation cannot meet the need for TPA, and the drug costs thousands of dollars a gram. Researchers have recently succeeded in introducing the gene that codes for TPA into the DNA of a goat, with the result that the animal produces TPA that can be isolated from its milk. In this process, known as "pharming," one goat can make as much TPA as a one-thousand-liter bioreactor.

Biotechnology may even prevent heart attacks from occurring in the first place. The little Italian hamlet of Limone Sur Grada has become famous because its inhabitants are free of heart disease, despite their high blood cholesterol levels. They have inherited a gene that codes for apolipoprotein A1, a special protein that scavenges cholesterol from the bloodstream. Injections of a genetically engineered form of this protein have dramatically reduced the clogging of coronary arteries in rabbits, a treatment that may eventually be viable for humans as well.

One day, perhaps, our young biotechnologist-to-be will work on this problem. But the day I encountered him, he was content to scrutinize a display about lysozyme, a natural milk enzyme with antimicrobial properties. We can use genetic engineering techniques to increase the levels of this enzyme in milk, thereby reducing udder infections and the need to dose infected animals with antibiotics. As the little guy wandered off, I noted that he stuffed his DNA sample into the back pocket of his jeans. Jeans that may have been dyed with indigo produced by recombinant

DNA and made of cotton genetically engineered to repel insects without the need for pesticides. Of course, not everyone shares my optimistic view of biotechnology. And, certainly, there are some controversial issues involved. So read on and we'll explore these.

FRANKENFUROR

The ancient Greeks did not have a good grasp of genetics. A giraffe, they thought, was a cross between a camel and a leopard, and an ostrich was the result of a camel mating with a sparrow. A tough task for the bird, one would think. Why did they hold such beliefs? Because, in the absence of facts, their imaginations took over. And they still prevail. A recent survey showed that a third of all Europeans believe that only genetically engineered tomatoes contain genes. Otherwise, the fruits are "gene-free," and, presumably, "risk-free."

Researchers undertake such surveys to gauge public reactions to genetically modified foods. It's probably the hottest potato to crop up in the area of food safety since pasteurization was introduced in the early 1900s. Activists back then advised people to spurn the new process because it destroyed the nutritional qualities of the milk, and they even described the horrors that could arise from consuming "dead bacteria." But the truth is that live bacteria were the ones they should have been worrying about. Of course, there are still holdouts who promote raw milk. They can have it.

Today's bogeyman is not pasteurization but genetic modification. Just about everyone has an opinion on the subject, but much too often people base their opinions on hearsay and emotion rather than on scientific data. Consumers speak of "Frankenfoods," and activists attack and destroy experimental fields planted with modified crops while at the same time they clamor for more research into the effects of such crops.

I am not going to suggest that there aren't some contentious issues surrounding genetic modification. There are, just as there are with any new technology. And I'm certainly not going to say that scientists can absolutely guarantee that there are no pitfalls involved in the genetic modification of foods. Nobody can offer such a guarantee. Indeed, those who demand unqualified assurance about the safety of genetically modified foods are just plain naïve. We don't make such demands concerning other aspects of life. We don't refuse to fly unless someone assures us that the plane will not crash; that would be absurd. We fly because we know that the benefits outweigh the risks. This is how we have to look at genetically modified foods as well.

Before delving further into the issues, we must acquire some understanding of what genetically modified foods are all about. Although the term "genetic modification" conjures up images of high-tech laboratories, humans have actually been modifying the genetic makeup of foods since time immemorial. If early farmers had not sprinkled pollen from one type of corn on another, we would still be looking at fifteen kernels per ear. If wheat had not cross-pollinated with some wild grasses, we'd have to contend with lower crop yields and more fungal damage. Without crossbreeding, we'd have no nectarines, seedless grapes, tangelos, or Mackintosh apples. We wouldn't even have grapefruit. This fruit first appeared in the eighteenth century, a result of the long-term crossbreeding of various citrus fruits. In each of these cases, genes from different species intermingled to bring out new, desirable traits. But this kind of genetic manipulation takes a long time, and it can sometimes foster undesirable traits.

Then, in 1974, scientists made a breakthrough. For the first time, they isolated and copied genes — those little segments of DNA molecules, found in the nucleus of every cell, that direct

an organism to carry out its myriad functions. In other words, they cloned them. This created the potential for inserting genes into the DNA of a target cell. Here the genes would be incorporated into the cell's genetic machinery and direct the cell to carry out some desired function. The scope of possibilities seemed almost unlimited. We could now guide plants to synthesize the insecticidal proteins or enzymes critical for the formation of natural anticancer substances.

Technically, plant genetic engineering is a complex business. The most common method makes use of a soil bacterium called *agrobacterium tumefaciens*. This bacterium contains rings of DNA, called plasmids, which we can remove and open up using specific enzymes. We can now add a segment of DNA from another species — that is, a gene; next, using another set of enzymes, we incorporate it into the plasmid; finally, we reintroduce it into the bacterium. This, in turn, we place in a solution along with a leaf from the plant that is to receive the gene. Here the bacterium infects the plant and transfers DNA from the altered plasmid into the plant's chromosomes. Chromosomes are the strands of DNA located in a cell's nucleus that are responsible for an organism's genetic makeup; they reproduce as a cell divides. The plant then grows with the new gene incorporated into its DNA, ready to express its desired trait.

This, then, is the technology that biotech companies are pursuing. And let's understand that just because something is good for Monsanto, Novartis, AstraZeneca, or any other company involved in biotechnology, it isn't necessarily bad for the public. But if you listen to the alarmists, you may form the impression that these companies are trying to foist poisons on us purely for the sake of profit. Naturally, there is a buck to be made. But profits come with the production of good and useful products. No company wants to undermine its existence by marketing dangerous substances. The industry has commissioned a great deal of research into genetic modification and its safety aspects. And we are seeing the practical benefits.

Genetic modification of crops to afford protection against certain insects is already well established. *Bacillus thuringiensis* is a natural soil bacterium that produces a protein, commonly referred to as Bt, which is toxic to many caterpillars. We can transfer the gene that codes for this protein into a plant cell's nucleus, and the plant will then thrive, even in the face of an insect infestation. Growers need to use fewer insecticides. In the U.S., four out of every ten pounds of insecticides used go to protect cotton, but farmers are already reporting that they have reduced their insecticide use by several million pounds annually. Canadian farmers who grow canola engineered to be resistant to the herbicide glyphosate (Roundup) report reduced chemical use. Farmers also till less often, which means less water pollution and less erosion. In China, Bt cotton is a big success story; more than two million farmers cultivate it. Production costs have dropped by twenty-eight percent, and the farmers' average annual income has increased. Use of toxic pesticides such as organophosphates has plummeted by eighty percent, and reported cases of pesticide poisoning among cotton farmers have decreased from twenty-two percent to five percent. Sounds good. So why are protesters dumping transgenic soybeans on the doorstep of Tony Blair, the British prime minister?

One of the concerns they raise is over insertion of a gene that makes a plant resistant to the antibiotic kanamycin. Scientists will sometimes incorporate this gene along with a set of desired genes as a marker to indicate whether they have successfully implanted the desired genes. If the plant is unaffected when placed in an antibiotic solution, then it is properly expressing the new genes. The question is whether these antibiotic-resistant genes are transferred to animals and to humans, decreasing the effectiveness of antibiotic drug treatments. (Actually, the antibiotics in question are not important for human use.)

Another worry is that pollen from genetically engineered plants may cross with wild plants, producing "superweeds." This has not surfaced as a problem yet, but pollen from genetically modified canola has spread more extensively than we had predicted. Farmers in western Canada who had no intention of growing modified canola now find that their fields are contaminated with it. Insects may develop resistance to Bt protein, giving rise to "superinsects," although experiments have shown that when we sow Bt-producing plants along with regular plants, we dramatically reduce such resistance. Of course, insects can develop resistance to any type of insecticide. Pollen from Bt corn may travel to other plants and produce unexpected effects. Cornell University researchers discovered in pilot studies that by sprinkling pollen from Bt corn onto the leaves of the milkweed plant they could kill the caterpillars of the Monarch butterfly, which dine exclusively on these leaves. But field trials have not corroborated these findings. Allergens may be inadvertently introduced. What are the consequences of introducing an antifreeze gene from the Arctic flounder into tomatoes to enhance frost tolerance? Will it present a problem for people with fish allergies if this technology is ever realized?

Many of the potential problems, such as the allergy issue, that are now being vocalized by opponents of genetic modification were, in fact, addressed long ago by the industry. Researchers have been testing for allergens in modified foods since the inception of the technology. In one case, the addition of a Brazil nut gene to soybeans in order to increase the quality of the protein in animal feed resulted in the transfer of an allergen. In other words, someone with a Brazil nut allergy could have reacted to eating the genetically modified soybeans. But researchers picked up the problem in routine testing, and the soybeans, which had only been destined for animal feed anyway, were never marketed.

This is quite a different approach from the one we take towards foods that have not been genetically modified. We don't ban peanuts, or strawberries, or fish because some people are allergic to these foods. And these allergies are far more prevalent than the theoretical allergies to modified foods. Indeed, we may be able to modify peanuts genetically in order to eliminate the protein that is responsible for allergies.

Opponents of genetic modification suggest that we should be satisfied with the normal process of crossbreeding plants to produce improved varieties. But where is the guarantee that this procedure doesn't introduce undesired chemicals? Appropriate crossbreeding can, for example, yield plants that are more resistant to insects. And why don't insects attack these plants? Because they contain more natural toxins than other plants. Nobody knows the human consequences of eating these natural pesticides. Why are the activists not demanding that all hybrid plants — or, indeed, that all plant foods — be tested for natural toxins?

Let me allow for the possibility that genetically altered foods present an as-yet-unidentified risk. One can always conjure up some theoretical catastrophe. But let's compare this to the very real benefits that genetic modification can offer. Combating malnutrition, for one. When people think of malnutrition, they usually think of starving children. But that is not the only kind of malnutrition out there. In fact, the most common kind of malnutrition in the world is iron deficiency. This can cause intellectual impairment, suppressed immunity, and complications in pregnancy. Millions suffer from iron-deficiency anemia. Most of them subsist on rice, a grain that contains very little iron, and the body cannot absorb the iron it does contain because of the presence of substances called phytates. These compounds bind iron in the digestive tract and substantially prevent it from being transported across the intestinal wall into the bloodstream.

Genetic modification has yielded a variety of rice that contains more iron. Scientists accomplished this by inserting a gene isolated from the French bean into the DNA of the rice. This particular gene codes for the synthesis of a protein called ferritin, which is an iron-storage protein. In other words, the new rice can take in more iron from the soil. Furthermore, scientists also added another gene — this time from a fungus — which codes for an enzyme that breaks down phytates, making iron more readily available.

Populations that subsist on rice also suffer from vitamin A deficiency. That's because rice is very low in beta-carotene, the body's precursor for vitamin A. Deficiency of this vitamin is a major cause of blindness in the developing world; estimates indicate that some 250 million children have vitamin A levels low enough to cause impaired vision. Lack of vitamin A also predisposes a person to various cancers and skin problems.

Researchers addressed this problem by introducing into rice four genes that code for the proteins that enhance beta-carotene synthesis: two from daffodils, and two from a bacterium. The rice they created was yellow, clearly demonstrating that it was now fortified with beta-carotene. Experiments are under way to cross the iron-rich rice with the beta-carotene-rich rice to produce a super-rice that will alleviate nutritional problems affecting billions of people.

We face many other fascinating possibilities. How about genetically modifying foods to contain higher levels of cancer-fighting compounds, such as sulphoraphane, found in broccoli? Or developing fresh fruits and vegetables with improved shelf lives? Or crops that will flourish in salty soil? What about tomatoes that not only taste better but also contain more anticarcinogenic lycopene? Seed oils with a healthier profile of fats are a possibility. So are crops that are more resistant to frost. We can use genetically modified potatoes to produce proteins that can serve as vaccines against human disease. All of these are real possibilities. Granted, we're not likely to see these benefits tomorrow, or next week, or even next year. The Wright brothers' first flight provides an appropriate analogy. The spectacle of their rickety airplane bouncing along for a couple of hundred meters was not very impressive, but anyone who saw it, and possessed a little imagination, realized that it was only a matter of time before the airplane would revolutionize travel. People had to witness it in action before they could accept it; then, over time, the details could be worked out. So it is with genetic modification.

As you see, the argument, pro and con, goes back and forth. My view is that the pros are likely to outweigh the cons, and that the biotechnology community has reasonably addressed the problems as they have arisen; it has offered effective arguments against a potential apocalypse. Still, many experience a lingering trepidation about consequences that we haven't yet considered. Let's face it — the intermingling of genes can lead to surprises. Parents who have crossed their genes to produce progeny can attest to that. But equating biotechnology with Chernobyl or thalidomide, as some do, is totally unreasonable. Admittedly, as we pursue biotechnology, we may cause some problems; but if we don't pursue it, we will likely encounter bigger dilemmas. Progress always comes at a cost, but if we fear the unknown, we will never get anywhere. Nothing in life is risk-free. Certainly, in the area of genetically engineered foods we do have to move cautiously and intelligently. This means that we must not rush headlong towards developing products that we do not want or need. It also means that we should not create alarm by dressing up in masks and contamination suits to handle transgenic soybeans, as some activists have done.

Genetic modification is a hugely complex scientific, economic, political, and emotional issue. And this is certainly not my last word on the subject. Maybe I'll even have to eat crow someday. But by then we'll probably have a genetically modified version that is nutrient-filled and highly palatable.

THE SUNSHINE VITAMIN

Tiny Tim, the young hero of Dickens's *A Christmas Carol*, probably suffered from rickets. In England, an epidemic of this crippling disease arrived on the heels of the Industrial Revolution, and it caused widespread misery. Children were most commonly affected. Many were forced to hobble on crutches because the weakened bones in their legs could not support them. Nobody at the time realized that this devastating affliction was linked to the newly built factories that spewed soot and smoke into the air. The skies over England were shrouded in a haze; the sun was often obscured. And that was the problem. As we learned much later, we need sunlight to produce vitamin D, a substance that is critical for proper bone formation. Furthermore, as recent research indicates, vitamin D may do more than strengthen our bones — it may also play a role in reducing the risk of some common cancers.

Vitamin D is often called "the sunshine vitamin." This is somewhat of a misnomer, because light is just a form of electromagnetic radiation and does not transmit vitamins. What it does do, though, is trigger the formation of vitamin D from its precursor in the skin. A series of experiments that researchers conducted in the early 1900s elegantly demonstrated the phenomenon. Dogs raised exclusively indoors developed rickets, but their condition reversed when the researchers exposed them to sunlight. It was a seminal experiment with rats, however, that truly clarified the situation. In the 1920s, Hess and Weinstock at Columbia University induced rickets in the rodents by depriving them of light. Then they excised a piece of skin from each and subjected it to sunlight. When they added the skin to the rats' food, the rodents rapidly recovered from rickets. Obviously, the light had converted some substance in the skin to an active form. Around the same time, Sir Edward Mellanby found another way of curing rickets. He, too, raised dogs in the dark until they had clearly developed weak bones. Then he tried adding various components to their diet. When he supplemented their feed with cod liver oil, the animals' bones normalized.

It was several years before scientists managed to work out the subtleties of vitamin D. We now know that its main role is to regulate blood levels of calcium by stimulating the formation of proteins that transport calcium across the intestinal wall. But this happens in a pretty complex fashion. The active compound has the foreboding name of 1,25-dihydroxy vitamin D_3, and it is formed in the kidney from its precursor, 25-hydroxy vitamin D_3, which in turn is formed in the liver from vitamin D_3. The latter compound is the one we find in certain foods, such as fish oils; but, more importantly, skin that is exposed to the sun forms the compound.

Once scientists understood this, a method of reducing the risk of rickets became apparent to them. If they could somehow produce vitamin D_3, then people could use it as a dietary supplement — or, even better, manufacturers could add it to some widely consumed food. The method of production the scientists developed was ingenious. They exposed skin from cows, pigs, or sheep to sunlight, and they extracted the vitamin D_3 that formed with a solvent. Milk, they decided, was the ideal vehicle for supplementation because it was a popular beverage, and it already contained calcium. Fortification of milk in North America began in the 1940s, and it soon reduced the incidence of rickets by a whopping eighty-five percent.

Over time, as childhood rickets has become more and more rare, attention has shifted to adults. More recent studies show that as much as sixty percent of the population have blood levels of vitamin D low enough to increase the risk of osteomalacia — also referred to as "adult rickets" — and osteoporosis, an even more serious condition. To a large extent, this is a result of the massive amount of publicity given to the link between sunlight and skin cancer. The elderly, especially, have taken to avoiding the sun like the plague. Too bad. Just fifteen minutes of sun exposure three times a week can dramatically increase vitamin D production and reduce the risk of fractures. Sun exposure is more effective than taking supplements; but, during winter in northern climates, the effective wavelengths of sunlight do not penetrate the atmosphere, and people must consider taking supplements. The usual recommendation is that those under fifty should have at least 200 international units (IU) of vitamin D_3 daily (the amount in two glasses of milk); 400 IU are appropriate for people between the ages of fifty and seventy; and people over seventy need 600 IU. Some researchers think that this is not enough, because surveys show that many people who take supplements still have low blood levels of 25-hydroxy vitamin D_3, and they recommend a daily intake of 1,000 IU. When physicians diagnose serious vitamin D deficiency, they usually recommend a simple remedy: 50,000 IU once a week for eight weeks.

Increased blood levels of vitamin D_3 may have other benefits. Breast, prostate, and colon cancers are more common in northern climates, possibly due to the fact that the inhabitants of these zones get less sun exposure. Some intriguing preliminary studies show that 1,25-dihydroxy vitamin D_3 can regulate cell proliferation and reduce the risk of these cancers. All of this has prompted me to look at vitamin D with renewed interest. I was never partial to this vitamin because of the horrors it conjured up in my mind. I still shudder when I think of how I was forced to take cod liver oil as a child. I remember resisting vigorously when my mother tried to force the foul liquid down my throat. Now I know that she wasn't trying to torture me — she was trying to keep me healthy, and she was succeeding in more ways than she knew. But I'm still not ready to swallow cod liver oil. A walk in the sunshine is a much more pleasant prospect.

That's the Way the Cookie Crumbles

I thought I had done a pretty slick job of hiding the cream-filled cookies in my shopping cart under a couple of bags of fruit. But I didn't get away with it. "You're not really going to eat those, are you?" a voice behind me asked incredulously. "They're filled with hydrogenated fat! Isn't that what you told us in class?" The gig was up. I had been caught red-handed by one of my own students. Sheepishly, I explained that I don't make a habit of buying cream-filled cookies. But when I saw these, I had a sudden urge to recapture some happy childhood moments spent dissecting the chocolatey cookies and licking the creamy filling. Now, that was fun! Who cared that the glorious filling was nothing more than sweetened fat?

As I got older, and somewhat wiser, I did begin to suspect that something that tasted so good couldn't possibly be good for you. And, sure enough, I learned that dipping into the cookie jar on a regular basis is not great for your cholesterol levels. Happiness comes at a price. That, one might say, is the way the cookie crumbles.

Cookie crumbling is actually a complex business. The extent to which it happens depends on several factors. Of prime importance is the development of gluten in the flour. Gluten is a three-dimensional network of protein molecules that forms when we knead dough with water. By adding water, we cause the proteins in the flour to unravel from their natural coiled position and form cross-links to each other. This sets up the molecular scaffold that supports the other ingredients. If the scaffold is strong, with many connections, the resulting texture is tough. High-gluten flour is great for bread, because for that we need a sturdy texture, but it is not suitable for cakes or cookies.

All-purpose flour has a lower protein content and therefore less potential for gluten formation. It's great for cookies. But the extent of gluten development is not solely a function of the protein content of the flour. The amount of fat, sugar, and water in the mix is also important. So is the mixing technique.

Fat tends to interfere with the ability of gluten to form a tough mass. It "shortens" the effect of gluten, coating the flour particles and physically separating them. Protein molecules within each particle can still form their three-dimensional gluten

network, but they cannot span the gap between flour particles. This results in a texture that is ideal for pastry and cookies. But even the type of fat we select has an effect. Making a cookie crumble right requires just the right amount of saturated fat. If you want a creamy filling, you'll also require saturated fat. And that's the problem.

So what is saturated fat? Think of it this way. Fats are made of a framework of carbon and oxygen atoms, with hydrogen atoms attached to the carbon chains. A fat that contains as many hydrogen atoms as the carbon skeleton can support is referred to as being "saturated" with hydrogen. In the case of unsaturated fat, instead of joining to hydrogen atoms, some of the carbon atoms forge additional linkages to each other. We call these fats "unsaturated" because they contain less than their full complement of hydrogen atoms. The shape of these molecules is also different. They are decidedly kinky at the position of the missing hydrogens. The molecules are bent, and they cannot be packed together as closely as the straight saturated fats. Closely packed fats are what make for crumbly cookies and flaky pastry. They also make for clogged arteries.

When consumers got wind of this modern plague, they mounted a campaign to pressure the food-processing industry to replace saturated fats with unsaturated ones. They wanted manufacturers to banish lard, butter, palm oil, and coconut oil and replace them with unsaturated fats derived from vegetable sources. Soon, labels on processed-food packaging proclaimed that the contents were "low in saturated fats," and hamburger emporiums announced that their french fries were now fried in "one-hundred-percent vegetable oil." Apparently, consumer pressure had curtailed the heart-damaging onslaught of saturated fat.

But food processors had long insisted that their cookies wouldn't crumble properly if they made them with vegetable oil, and that unsaturated vegetable fats did not meet the requirements for high-volume fast-food frying. Then, all of a sudden, these major technical difficulties appeared to be licked; suddenly, we could relax and dine on fatty cookies and french fries, our blood gushing freely through our clean arteries. Well, we shouldn't have allowed ourselves to relax quite so fast. While the food-processing industry did, for the most part, switch from saturated to unsaturated fat, we must bear in mind that all unsaturated fats are not the same. Some actually behave like saturated fats in the body. Here's the story. We can solidify an unsaturated vegetable oil so that it will behave more like a saturated fat — that is, we can "partially hydrogenate" it. Treatment with hydrogen gas allows some hydrogen atoms to be inserted into the molecule. Unfortunately, not only does this process make the fat more saturated, but it also converts some of the unsaturated fat molecules into a slightly different, although still unsaturated, form. These so-called trans-fatty acids have had the "molecular kink" taken out of them, and their long straight chains can now cluster together, behaving just like the infamous saturated fats we use in cookies and fried foods.

So, in a sense, we have leapt out of the frying pan and into the fire. Consumers may gain confidence by reading labels assuring them that a food contains no saturated fats, but that confidence wavers when they are confronted with the issue of trans-fatty acids. The bottom line is that trans-fatty acids, which on a product label can fall under the "unsaturated" umbrella, may be just as damaging to arteries as the notorious saturated fats. They may have taken the kink out of the molecule, but the hype about reduced saturated fats is still pretty kinky. And what happened to my desire to recapture my childhood through cookies? I tried those cream-filled concoctions. They were not as good as I remembered. So, I guess I'll stick to fruit for snacks. No hydrogenated fats there. And I don't have to hide fruit in my shopping cart.

Coenzyme Q10 Is Worth Remembering

The gentleman in the front row was waving his hand energetically. He obviously had something to contribute to the discussion. I had just finished a public lecture on dietary supplements, and I threw the floor open for questions, anticipating the usual queries about glucosamine, calcium, vitamins, and the myriad herbal concoctions that have recently flooded the marketplace. But the waving man, who was looking pretty spry for his age, didn't have a question. He had a story to tell. A dietary supplement had changed his life.

Until a few weeks earlier, he had been in a pretty sad state. Diabetes and a heart condition had robbed him of his vigor, making him practically immobile. Then he heard about a pill that could work wonders. I asked him to tell us what this magical pill was, expecting him to name one of the numerous multilevel-marketed "natural products" that, their makers claim, relieve every ailment known to humankind. Such panaceas are not recognized by the scientific establishment, the story usually goes, because if they were then prescription drug sales would suffer. When you're in the business of delivering public lectures, you often hear such accounts. They come complete with testimonials about miraculous recoveries and heightened feelings of well-being. Much of this we can ascribe to the power of suggestion and people's unfamiliarity with the nature of disease. Many conditions would resolve themselves without intervention, but if some intervention has been made, it gets the credit.

I was therefore fully prepared to hear about super blue-green algae, shark cartilage, Rubyasia from the Peruvian rain forest, "natural clay" tablets, or magnetized-urine capsules. Unfortunately, though, the gentleman could not remember what miraculous substance he had taken to turn his life around, other than that it was in the form of a little yellow pill. He offered to go to a nearby pharmacy and return with the information. Thinking that this was the end of this little escapade, I went on to answer the usual questions about vitamin E, ginkgo biloba, and milk thistle. And then suddenly the gentleman was back. He had scampered off to the pharmacy and returned, all with impressive speed, clutching a piece of paper with "coenzyme Q10" written on it. That sure got my attention.

This substance had long intrigued me, and I'd always wondered why the medical establishment had not paid more attention to it. It's not your usual quack supplement; it has some pretty solid research behind it. In 1957, Dr. Frederick Crane at the University of Wisconsin became interested in how the energy needed to power the heart was produced. Using a beef heart as a model, he managed to isolate a substance that seemed critical to this process. Specifically, coenzyme Q10 is involved in the complex series of reactions by which carbon and hydrogen atoms from food combine with oxygen to form carbon dioxide and water, releasing energy. The body exploits this energy to synthesize a marvelous molecule called ATP (adenosine triphosphate), which then stores the energy. When the body needs to move a muscle, heat itself, or power the heart, ATP undergoes a chemical reaction that liberates the required energy.

Such chemical reactions involve the reorganization of the electrons that glue atoms together to form molecules. This reorganization, in turn, involves a type of electron transport system in which the needed electrons are passed from one molecule to another until they arrive at their destination. Coenzyme Q10 is a critical molecule in this transport system. Without it, the body could not produce ATP and supply energy. Crane's discovery generated a great deal of interest, because its potential application was pretty obvious. If we could increase coenzyme Q10 levels in cells, then ATP synthesis would be enhanced and energy production increased. It took no more than a year for Karl Folkers at the University of Texas to determine the molecular structure of CoQ10, synthesize it, and begin studying its properties.

On top of its electron-transport ability, CoQ10 turned out to be a potent antioxidant, protecting cell membranes from being destroyed by oxygen. This was enough to inspire supplement manufacturers to put the cart before the horse, and they started hyping CoQ10 as a miracle drug. They claimed that their product could provide relief from heart disease, cancer, aging, immune problems, and poor exercise tolerance. This is just the kind of activity that sours the scientific community — researchers were quite aware that studies had not corroborated such outlandish claims. As a result, many scientists tossed CoQ10 on that great junk heap of supplements that fail to live up to their promise.

A handful of researchers, however, struggled on with CoQ10 studies. They were impressed by the fact that blood levels decline with age, and that the heart has the highest concentration of CoQ10 in the body. They organized trials with patients suffering from various heart ailments, particularly congestive heart disease. (Of course, the subjects continued to receive traditional treatment as well.) Indeed, nine placebo-controlled trials have confirmed the effectiveness of CoQ10 in reducing shortness of breath, hypertension, palpitations, and chest pain. Echocardiographic measurements have shown improved heart function. And, perhaps most important, no study has brought to light any detrimental effects, even at a few hundred milligrams a day. Strangely, the medical community remains largely unaware of the CoQ10 research, but perhaps it's because the pharmaceutical companies have not promoted it. The substance cannot be patented, so no one has subjected it to a marketing push, in spite of the fact that CoQ10 may reduce insulin requirements in diabetics, help heal gum tissue in gingivitis sufferers, and even, in the form of a cream, improve skin condition. Another interesting observation is that CoQ10 levels are lower in people taking the statin drugs to reduce cholesterol. Perhaps we will eventually see a recommendation that patients taking these drugs also take a CoQ10 supplement.

Certainly, not every CoQ10 study has demonstrated the substance's benefits, but enough have to justify further investigation. I'm not ready to dismiss as placebo effects the benefits experienced by the gentleman who didn't even remember the name of CoQ10. It may not improve memory, but CoQ10 is worth remembering.

STICKING IN YOUR MEMORY

Some things just stick in your memory. When I was in elementary school, a teacher attempted to dissuade us from chewing gum in class with the following ditty: "The gum-chewing student and the cud-chewing cow are different somehow / I think the difference is the intelligent look on the face of the cow!" The message was an effective one, because it did deter me from exercising my jaw in that particular fashion. To this day, whenever I'm offered a stick of gum, the image of the cow springs to my mind, and I usually politely decline. I cannot remember the name of the teacher who was responsible for this life-changing recitation. That bothers me, so I've decided to do something about it. I've decided to take up gum chewing. But not ordinary gum — Brain Gum.

I came across this product when I was doing some research into the chemistry of memory. It all started one morning when a fly decided to investigate the cornflakes in my cereal bowl. I shooed him away, but he soon returned. Once more, I evicted him, but after buzzing around a little he again landed on the rim of the bowl. Our fly-and-man game went through another four or five cycles before something remarkable happened. The fly transferred his attention to another cereal bowl on the table. He landed, took off, landed, took off, but he never came back to my bowl again. The insect had apparently learned that my bowl presented dangers that the other one did not. I had never considered the possibility that flies could be trained. Had I discovered something here? Alas, as it turns out, no. A search of the literature revealed that others have indeed investigated fly training.

Researchers at Cold Spring Harbor Laboratory on Long Island found that they could condition fruit flies. Flies placed in a licorice-scented container where they received an electric shock learned to avoid the scent of licorice. But not all flies

learned the same way. Some needed to be zapped ten times before they avoided the licorice scent, and some never learned. The researchers then bred the smart and the dull flies separately, creating two genetically altered strains. The offspring of the smart fruit flies showed remarkable memory. Many learned to avoid licorice scent after being zapped just once.

What was going on here? When the researchers studied the flies' brains, they discovered major differences. They found more connections between the nerve cells of the smart flies than between those of the dim-witted ones. I'm not sure how one actually investigates the neurology of fly brains, but I assume that if we can go to the moon and invent Twinkies, then we can do it — in a pretty detailed fashion. Apparently, the formation of the links between nerve cells is controlled by a protein that turns on the genes that send out the message "Make connections to other nerve cells." The flies that never learned their lesson had very low levels of this protein, while the quick learners were rich in it. And guess what. The researchers now say they have hit upon chemicals that can stimulate production of the "smart" protein. Sounds good, but does the world really need superflies who may remember who tried to swat them and then use the information to plot revenge? No. But the interesting thing is that human memory seems to depend on the same kind of connections between brain cells.

The prevailing view used to be that as we age, our brain cells die, and we cannot replace them. Recent research, however, shows that the brain continually rewires itself as new cells are spawned and go on to make connections with others. Somehow, the mysteries of memory are locked up in these intercellular connections. The first evidence we had of this came from mice. Test mice that had running wheels in their cages learned to navigate mazes more quickly. Researchers found that their brains possessed more newly formed nerve cells. Most of us wouldn't consider running on a wheel intellectually demanding, but it's evidently challenging enough for mice to improve their brain functions. Mice only have to run mazes, but London taxi drivers have to navigate the most complex network of streets in existence. It's a mind-boggling challenge. What is happening inside their brains as they learn to whisk their customers efficiently from one point to another? Believe it or not, their brains are expanding — at least, the part of the brain responsible for storing and retrieving memories is expanding. Magnetic resonance imaging (MRI) scans of London taxi drivers in training have shown that their hippocampus grew during the two years they spent learning their way around the city.

Most of us would like to improve our brain functions, but not necessarily by driving a taxi around London. What are we to do? Getting an education seems to help. The more we use our brains, the more connections are forged between our brain cells. Some studies have even suggested that people with less education have a higher risk of Alzheimer's disease. Intellectual activities — such as solving puzzles, playing chess, and trying to figure out how Brain Gum can improve memory — are particularly good for the brain.

So we are back to the Brain Gum. Another silly marketing gimmick designed to capitalize on people's fear of intellectual inadequacy? Maybe not. Brain Gum contains phosphatidylserine (PS), a compound that improved cognitive function in mature adults in double-blind studies. PS occurs naturally in the brain, and it is located in the membranes that surround every cell. It plays a mysterious but important role in allowing those all-important connections to be forged between brain cells, and it also makes cells more responsive to the neurotransmitter chemicals that brain cells require to communicate with each other.

A typical double-blind trial involved over four hundred elderly people with memory problems who were given three hundred milligrams of PS daily over six months. They showed significant memory improvement when researchers assessed them using several tests. The dose they took was roughly the dose one would get from chewing the recommended amount of Brain Gum. PS is derived from soybeans and is also available in pill form. Maybe it is through PS that those genes turned on by the protein in the fruit fly experiments carry out their work. Some brainy scientist who has plenty of PS in his cell membranes will figure it out one day.

Memory is a complex business, and no single supplement will work magic. Increased intake of B vitamins has been linked with memory improvement, as have antioxidants such as vitamins E and C. We also know that estrogen supplements in women foster nerve cell growth, increase branching of nerve cells, and help repair damaged cells. Physicians have noted that patients on estrogen experience improvements in memory and concentration. In the U.S., health food stores sell huperzine A (Cerebra), which is extracted from a Chinese moss, as a memory enhancer. It inhibits an enzyme, cholinesterase, which breaks down acetylcholine, a neurotransmitter involved in memory. Huperzine has not undergone proper clinical trials as yet, but early studies give us reason to be optimistic. Researchers have not yet evaluated its side effects. It may turn out to be more effective than the herbal extract ginkgo biloba, which, in spite of the hype, has little clinical evidence to support its use. Research has demonstrated, however, that physical exercise retards brain deterioration.

And what is the simplest way to improve memory in aging adults? Well, it just may be eating mashed potatoes or barley. University of Toronto researchers found that the memories of a group of elderly people who ate a serving of potatoes or barley — as compared with others who consumed a beverage containing no calories — improved for about an hour. The improvement was certainly more modest than that claimed by the makers of Brain Gum, who insist that Brain Gum chewers will experience improvement in name recall, improvement in remembering information, and improvement in finding misplaced objects. But leave it to scientists to throw cold water on a hot idea. The memory improvement may have nothing to do with PS — it may be initiated by chewing any old gum.

A joint study conducted by researchers at the University of Northumbria and the Cognitive Research Unit in Reading, England, found that people recalled more words and performed better in tests on working memory after chewing gum. The study's authors suggested that gum chewing delivers more oxygen to the brain. Hard to see how that would happen. Still, I decided that if I was to give gum chewing a try in an effort to improve my memory, I might as well try Brain Gum. So I purchased a pack, in spite of its hefty price. The trouble is I can't remember where I put it. Maybe I should go and make myself some mashed potatoes.

Take It with a Grain of Salt

The processed food industry loves salt. Sodium chloride is cheap, allows water to be retained, acts as a preservative, and enhances flavor. As one salt promoter says, "Salt is what makes things taste bad when it isn't in them." True enough. We may partially explain our craving for salt by attributing it to our physiological need for sodium. Without it, our nerve cells can't transmit electrical impulses, our muscles can't contract properly, and our body fluids go out of kilter. So it shouldn't come as a surprise that "salty" is one of the basic human tastes. But salt does more than just add saltiness to food; it can also modify the way we perceive the other common tastes — namely, sour, bitter, and sweet.

Salt inhibits bitterness, and it can enhance sweetness. That's why you'll find salt in such unlikely foods as chocolate, apple pie, and breakfast cereals. Indeed, studies have shown that consumer acceptance drops dramatically when salt levels in processed foods decline. That would explain the popularity of foods such as dill pickles, hot dogs, sauerkraut, vegetable juices, cottage cheese, olives, canned soups, and pizza, which can have up to a gram of salt per serving. It isn't hard to see how one can easily exceed the recommended intake of six grams a day.

Salt was the first seasoning our ancestors used. They got it by evaporating seawater or by mining it. We have traced the origin of salt deposits in the ground back to oceans that no longer exist; essentially, all salt is sea salt. People mined salt near Salzburg ("City of Salt") in Austria as early as 6500 B.C., and the ancient Romans built large evaporation ponds by the sea to collect the stuff. In fact, the Romans valued salt so highly that they gave their soldiers a special allowance, known as the "salarium," to purchase it. (Our word *salary* derives from this Latin expression.) Salt was so important that many believed the person who spilled it would attract bad luck and malevolent spirits. Tossing a little salt over the shoulder was the antidote: the grains of salt would lodge in the spirit's eyes and distract it from the evil it was planning. Spilled salt as an omen of bad things to come was an enduring belief. In Da Vinci's painting *The Last Supper*, we clearly see an overturned salt container in front of Judas, foreshadowing his betrayal of Jesus.

It wasn't only for its taste that salt was so prized. It was also for its value as a preservative. When the salt concentration outside a bacterial or fungal cell is higher than it is inside, water is drawn out of the cell to reduce the outside salt concentration. This process of osmosis dehydrates the cell and eventually destroys it. That's why our ancestors rubbed salt into wounds to reduce the risk of bacterial infection. But this would also disturb tissue cells and cause the irritation we associate with "rubbing salt into the wound." At one time, people also preserved meat by soaking it in a brine solution or by covering it with whole grains of salt (which were known as "corn," hence the origin of "corned beef"). Perhaps the most unusual use of salt as a preservative was devised by the seventeenth-century British authorities who put the heads of executed villains on public display to deter other criminals. They discovered that the heads would rot quickly and attract birds; the birds would strip off the flesh, leaving behind a clean skull, which was apparently less frightening. The answer to this little problem was to boil the heads in salt water so they would not putrefy.

These rogues were salted after death. But what about the possibility of salt bringing on death? Our bodies try to maintain a certain concentration of sodium in the blood. If the amount of sodium rises, the body will retain more water in order to maintain the same concentration. This increases the blood volume, so there is more blood for the heart to pump around the body. The pressure the blood exerts against the walls of the arteries intensifies, and this can lead to stroke or heart attack. But, if the body takes in less sodium, then it will retain less water, and the blood pressure should go down. "Go easy on the salt" is the advice of the physician who has diagnosed high blood pressure.

Numerous studies have shown us that about fifty percent of those suffering from high blood pressure respond to a low-sodium diet. Why not all of them? Because controlling blood pressure is more complicated than simply striking a balance between sodium and water. Calcium and potassium play important roles as well. In fact, many researchers now believe that increasing potassium and calcium intake is as important as reducing sodium intake for people with high blood pressure. This means more skim milk, more bananas, and more oranges.

While no one disputes that the low-sodium diet is an important treatment for people with high blood pressure, experts bicker when it comes to making recommendations for the public at large. Some say the argument that everyone should reduce their salt intake from about nine to six grams a day is not based on science. I think they are wrong. Many people have undiagnosed high blood pressure and would benefit from a reduced salt intake. Experiments with chimps have indicated that as salt in the diet increases, blood pressure rises. Human epidemiological studies demonstrate the same thing. Populations with low salt intake have lower blood pressure. The Yanomani Indians of Brazil add no salt to their food, and they do not develop hypertension — despite being surrounded by poisonous snakes, bugs, and researchers who constantly want to measure their blood pressure.

By contrast, we North Americans, with our penchant for salty hot dogs, chips, and pizza, are in the midst of a hypertension epidemic. Whether a reduced-salt diet lowers blood pressure in people who do not have high pressure to start with is irrelevant. Eating less salty processed food automatically translates to a healthier diet. I'm sure that spokespeople for the influential Salt Institute, an organization that promotes the use of salt, will dispute this. But I would take their comments with a grain of salt.

BOTULIN: DEADLY POISON AND FASCINATING MEDICINE

The young mother was surprised by the doctor's question. "What does your little girl eat for breakfast?" "Only hot oatmeal with milk," she answered. "Does she put any sugar on it?" the doctor queried. Now the lady became indignant. "We eat only whole, natural foods. No meat, no processed food, no sugar. Sugar is poison. We use only natural honey on the oatmeal." And with those words she confirmed the doctor's suspicion. The little girl's baby brother, whom they had brought to the hospital suffering from a mysterious ailment, had botulism poisoning.

It was a bizarre case. The three-month-old baby suddenly stopped nursing, and his body became progressively floppy. Within four days, he was practically lifeless, so they rushed him to emergency. At first, the doctors suspected spinal muscular atrophy, a rare neurological disease that is essentially a death warrant. But this horrific disease doesn't usually come on so suddenly. Such a rapid onset of symptoms smacks of poisoning. Botulism would explain the muscular flaccidity, but why wasn't the mother affected? She swore that she'd fed the baby nothing but breast milk. The doctor, however, was not so sure. That's what prompted him to ask those questions about the boy's sister. Yes, she did like to help with the baby, the mother divulged. Sometimes she even pretended to feed him with an empty spoon. Now the lights flashed in the physician's mind. When her parents weren't watching, the little girl probably did more than pretend, treating her baby brother to a bit of her honey-laced oatmeal. Unfortunately, the honey was likely laced with the poison produced by the *Clostridium botulinum* bacterium.

Laboratory tests confirmed the physician's suspicions by revealing the presence of the toxin in the baby's serum and feces, as well as in the jar of honey. The little boy eventually recovered, after spending five weeks on a respirator. That's what botulinum toxin can do to you — if you're lucky. Want to know how poisonous the substance is? Picture a grain of salt. Now imagine dividing this grain into roughly a million pieces. Each piece will weigh about one nanogram, the amount of botulinum toxin it takes to kill a human. How did the toxin get into the honey? Easily. Spores of this deadly bacterium are everywhere. They're in the earth, in the air, and in the pollen and the nectar that the bees gather. Spores are forms of the bacterium that exist in a sort of suspended state of animation. They do not feed, and they don't reproduce until they encounter the right conditions. Low acidity, moderate temperatures, and the absence of oxygen make them come alive and start spewing venom. Conditions in the gut are favorable to these venomous bugs, but there they encounter a problem: other bugs. The human gastrointestinal tract harbors numerous species of bacteria, which all compete for food; luckily for us, botulinum bacteria do not fare well in this contest, and they cannot establish themselves. Unless, of course, there are relatively few competing bacteria, as is the case with the gastrointestinal tracts of infants. That's why children younger than twelve months old should not be fed bee regurgitation — that is, honey.

While we adults don't have to worry about botulinum spores germinating in our digestive tracts, we do have to be concerned about eating food in which the bacteria have multiplied and produced their toxin. In Vancouver, thirty-seven people were poisoned when a restaurant used contaminated garlic to make garlic bread. Dirt that adheres to garlic bulbs commonly contains botulinum spores. If we store the cloves under oil in an anaerobic environment and keep the jar at room temperature, then the bacteria will come alive and produce the toxin. When this toxin enters the bloodstream, it binds irreversibly to nerve endings and prevents the release of acetylcholine, the neurotransmitter that triggers muscle activity. The result is droopy eyelids, double vision, difficulty speaking and swallowing, progressive weakness, and, finally, paralysis of the chest muscles and respiratory failure. If the victim survives the initial onslaught of the toxin, recovery is likely because the affected nerves will eventually sprout new branches capable of acetylcholine release.

Garlic in oil is not the only problem. The first account of botulism — recorded in 1822 by Justinus Kerner, a German physician — focused on sausages, which provide an internal environment conducive to the growth of the bacteria. Indeed, the term *botulism* was coined from the Latin word for sausage. Not to worry, though. Today, food processors add nitrites to sausages and hot dogs to prevent the growth of this nasty microbe. But another concern has cropped up. Oil infusions are now popular offerings in some restaurants. Adventurous chefs use vegetables, herbs, spices, and even lobster meat to flavor oil, which they then use to add a flavor dimension to a variety of dishes. If improperly prepared, these oils can be lethal. The chef must use only well-washed produce, add lemon juice or vinegar to the oil (one tablespoon per cup of oil), shake the mixture vigorously, store the infusion in the refrigerator, and keep it for no longer than a week. We can destroy botulinum toxin by boiling for ten minutes, but spores are resistant to heat. Only by cooking a contaminated food in a pressure cooker for fifteen minutes can we kill them. We are able to subject canned food to these conditions during processing, but not infusions.

Ah, I can practically see you frowning out there. But be careful. Frown too much and those wrinkles between your eyebrows could become permanent. There is a treatment, though. Ask a doctor to inject tiny amounts of botulin into the muscles of your forehead; this will paralyze them, making those wrinkles practically disappear. An ingenious use of a toxin, half a glass of which could kill the entire population of the world.

From the Jungle to the Operating Room

We are in the jungles of South America. It's the late sixteenth century. Monkeys are jumping from tree to tree. Suddenly, one of them emits a shrill cry. He manages to jump to another tree and then to one more before falling to the ground, dead, an arrow protruding from his side. The poison was not a particularly potent one. The natives call it a "three tree" poison — as opposed to stronger formulations, which would be termed "two tree" or, the ultimate, "one tree."

Sir Walter Raleigh himself witnessed such an event, and he was dumbstruck by the quickness of the monkey's death. He asked to examine the mixture that coated the arrow. Raleigh then took a speck of the "urari," as the stuff was called, and rubbed it between his fingers. Sir Walter must have had a small, unhealed cut on one of his fingers, because he immediately became dizzy and promptly collapsed as the poison entered his bloodstream. Luckily for him, it was only a three-tree dose.

Raleigh learned the meaning of "urari" the hard way. In the language of the natives, the word meant "he to whom it comes, falls." He also learned that some tribes put the poison under their fingernails, where it would come in handy during hand-to-hand combat. Stories circulated about how certain tribe members secretly mixed the lethal brews. The tribe's oldest women would perform the task in closed huts; if, after two days, the fumes had not killed them, the mixture would be judged too weak to use, and the women would begin work on another batch.

The active ingredient in the preparation turned out to be the root or stem of a certain species of climbing vine, known today as *Chondodendron tomentosum*. Sir Walter took a sample back to Europe, where it was given the name "curare," derived from the Indian word for poison. No one paid much attention to the substance until 1812, when Charles Waterton learned that if he administered the right dose, he could achieve muscle relaxation without death. Doctors began using curare to treat lockjaw, infantile paralysis, and even epilepsy.

It wasn't until 1844 that scientists started to acquire an understanding of the drug's mechanism of action. The French physiologist Claude Bernard experimented with frogs. He found that curare blocked nerve impulses from the brain to the muscles and had the effect of relaxing the muscles to the point of limpness. Even the muscles that controlled breathing could be made to relax to the point that the frog appeared to be dead. If the dose was just right, the effect would soon wear off, and recovery was complete. One might say that the frog had experienced a living death.

The concept of living death brought fear to the hearts of nineteenth-century Europeans and Americans. Edgar Allan Poe dramatized this fear in his classic tale "The Premature Burial." The story was written in 1844, just around the time that stories about curare's effects began to spread. Various accounts of individuals being mistakenly declared dead and waking up inside their coffins also made the rounds. People were scandalized by stories about unearthed coffins containing skeletons with hands clutching at the lid from the inside. Some went as far as having elaborate safety mechanisms built into their coffins so that they could sound the alarm should they find themselves prematurely buried.

Poe's tale of premature burial is haunting. The reader shares the protagonist's panic as he emerges from a condition called "catalepsy" and discovers that he's enclosed in a wooden container and enveloped in the smell of fresh earth. We hear his terrified screams and feel his horror as he realizes that his worst nightmare has come true. The screams, however, do bring help. Reassuring hands rouse him, and we learn that he has not, in fact, been buried alive; earlier, he'd taken refuge from a storm on a docked boat and fallen asleep in a rather cramped wooden bunk. The boat carried fertilizer, hence the smell. The burial was not real, but the panic was. Did knowledge of curare contribute to Poe's obsession with death and to his fear of premature burial? Very possibly, since scientists working with curare had recently shown that deathlike states really did exist.

When those scientists isolated the active ingredient in Sir Walter Raleigh's climbing vine and identified it as tubocuranine, modern medicine soon found applications for the substance. It could counter the effects of some muscle-contracting poisons, such as strychnine and tetanus toxin, and, even more importantly, it could serve as a muscle-relaxant for surgical patients. Curare greatly facilitated abdominal surgery by preventing the muscles from becoming stiff and almost impenetrable. In 1942, Dr. Harold Griffith, chairman of McGill University's Department of Anesthesia, became the first to attempt this application, using curare during an appendectomy at Montreal's Homeopathic Hospital (later the Queen Elizabeth Hospital).

Dr. Griffith spent his life researching anesthesia, and he probably deserves more credit than anyone else for establishing the field as a medical specialty. He received numerous awards for his work, but his contributions are best encapsulated in his biographer's comment: "There are only two eras in anesthesia, before Harold Griffith and after."

We rarely use curare these days because more effective synthetic derivatives have supplanted it. These — including pancuronium, better known as Pavulon — are more potent, have more limited side effects, and can be given in smaller doses. But in larger doses, Pavulon can cause paralysis and death, as illustrated by a notorious 1975 criminal case.

During a six-week period, thirty-five patients suffered fatal or near-fatal cardiopulmonary arrest in a veterans' hospital in Ann Arbor, Michigan. All of them had been on intravenous lines. The FBI investigated and found Pavulon in the tissues of the five patients who had died. The bureau charged two nurses, both from the Philippines, with murder and attempted murder. The prosecution claimed that the women were trying to dramatize the need for more nurses. Imelda Marcos, the first lady of the Philippines, paid for part of their defense, insisting that the FBI had trumped up the charges in order to keep foreign nurses out of the country. Wherever the truth lay, the nurses eventually served only a few months in jail. They were released when a mysterious judgment was issued claiming that irregularities had occurred during the trial. Some people say that Imelda had threatened to make trouble over American bases in the Philippines. Perhaps, in this case, curare paralyzed the long arm of the law.

The Growing Growth Hormone Industry

"I expect to see 150. I'll be disappointed if I don't." That remarkable statement comes from Dr. Ronald Klatz, president and founder of the American Academy of Anti-Aging Medicine, an umbrella organization of physicians dedicated to using treatments that may slow the aging process. Klatz is optimistic about his life expectancy because he regularly injects himself with human growth hormone. Thousands of other North Americans are also puncturing their thighs, abdomens, or bottoms daily with syringes filled with this purported antiaging miracle. And they are opening their wallets wide to do so. Those who cannot afford the fifteen hundred dollars a month for the hormone injections are banking instead on dietary supplements, known as "secretagogues," which allegedly stimulate the release of growth hormone in the body. But before these hopefuls start reserving cruise tickets to celebrate their hundredth wedding anniversaries (presuming they have enough money left after paying for all that growth hormone supplementation), I think they should take a closer look at this alleged shot of youth.

To be sure, human growth hormone (HGH) is a fascinating substance. If our pituitary gland doesn't release it properly, then we simply do not grow. "Pituitary dwarfs" remain short in stature, but they are otherwise normal. Today, we can treat children affected by this condition with injections of growth hormone. Before the late 1970s, the only source of this hormone was the pituitary of human cadavers. Extraction was difficult, and, in a few cases, tragedy occurred when viruses from the donor contaminated the sample. This is no longer a danger, because today we produce the substance using E. coli bacteria that have been modified with the gene that codes for human growth hormone production. Thanks to this technique, the hormone has become widely available, and the situation has stimulated research on the use of growth hormone for conditions other than lack of growth. One of these conditions is aging.

We experience a decline in growth hormone production as we age, as can be demonstrated by measuring hormone levels in the blood. Actually, HGH is difficult to measure, so we generally monitor it through levels of insulin-like growth factor 1 (IGF-1), which the liver cranks out in response to human growth hormone. IGF-1 is the hormone that carries out the work attributed to growth hormone. We measure it in blood plasma as units (U) per liter. Young men show values in the range of 500 to 1,500 U, with only five percent falling below 350 U. Of healthy men over sixty, however, thirty percent have values below 350 U. On the one hand, we could argue that this is to be expected, because we have no need for growth hormone once we have stopped growing. On the other hand, adults who have pituitary disease commonly show growth hormone deficiency and exhibit symptoms such as increased body fat, reduced strength, and impaired psychological well-being. Doctors can reverse these symptoms in their patients by administering growth hormone, so the question of what happens if growth hormone is given to aging men who are healthy but exhibit low blood levels of IGF-1 is certainly an appropriate one to consider.

Dr. Daniel Rudman of the Medical College of Wisconsin addressed this very question in a landmark study published in *The New England Journal of Medicine* in 1990, a study that triggered all the subsequent hype about HGH. Hype that was mostly unwarranted. First of all, the study was small. Only twelve healthy men between the ages of sixty-one and eighty were treated, and only for six months. Some of the results were certainly encouraging. Body fat decreased by fourteen percent, while lean body mass and skin thickness increased by nine percent and seven percent, respectively. We could interpret these as modest antiaging effects, but we have to remember that the researchers selected these subjects because they had low IGF-1 levels to start with. Two-thirds of the elderly do not fit into this category.

Furthermore, the fact that both blood pressure and plasma glucose levels increased in the experimental group was somehow lost amid all the glowing reports. Neither do most people who read the original study know that it continued after the preliminary results were published. They did not hear that several of the men developed carpal tunnel syndrome, a painful hand condition, or that a couple of them experienced breast growth. They may not have heard that the authors of a 1996 study published in *The Annals of Internal Medicine*, while confirming Rudman's results, found that in spite of the antiaging effects the subjects enjoyed no measurable improvements in muscle function, strength, or physical performance. They did, however, suffer an increase in joint pain and breast enlargement. Some researchers began to wonder whether growth hormone might also stimulate the growth of tumors in the body.

In spite of these caveats, clinics started offering HGH injections. Anecdotal reports of wondrous effects flooded in, unconfirmed by any controlled studies. Then the market expanded to snare those who could not afford the expensive injections but who might be willing to swallow cheaper dietary supplements. These people also had to swallow some questionable science. Such as the notion that certain blends of amino acids significantly increase growth hormone production and trigger rejuvenating effects. Many of the supplement promoters referred to the Rudman study, which had nothing to do with their supplements and gave no support to their claims. It is also worth noting that Dr. Rudman (now deceased) did not back the claims of the HGH promoters, and he did not use the hormones himself, even though he was in the age category of the subjects who participated in his famous study.

Perhaps further research will justify the use of HGH to treat the elderly who are deficient, and possibly even those who have normal levels. But for now, anyone who is contemplating jumping on the HGH bandwagon should first read a study published in *Nature*, a top scientific journal. In it, researchers at the University of North Dakota record their discovery that mice with a growth hormone deficiency lived longer than normal mice; they also reported that small breeds of dogs and horses have increased life expectancies. Perhaps we should examine how humans with reduced levels of growth hormone fare in terms of longevity. I'm sure that Dr. Klatz would reject this approach. He says that if his daily injections of HGH fail to make him live to 150, then he will be "disappointed." I'm betting he's in for a disappointment. A major one.

STRONG POISON

"The reason that people die from arsenic is because they believe it to be poisonous." At least, that's what Mary Baker Eddy, the founder of Christian Science, claimed after she had been introduced to homeopathy, the popular practice that supposedly cures people with solutions that are so dilute they essentially contain nothing. Eddy surmised that since the adherents of homeopathy were cured with nothing, then disease must exist only in the imagination. One could achieve cures without physical intervention. She was not completely wrong about this: the mind is a powerful force. But she was certainly wrong about arsenic.

Just a few years before Eddy uttered her confused remarks about arsenic toxicity, the English government had introduced the Food and Drug Adulteration Act of 1860, prompted by a tragic case that demonstrated that arsenic poisoning was not all in the mind. It was in the liver, the kidneys, the blood, and the skin. At the time, druggists would mix calcium sulfate (plaster of Paris) into peppermint lozenges as a whitening agent. One day, as an assistant was preparing a batch of the candies, he accidentally reached for the wrong powder. Arsenic oxide, which was sold as a rat poison, ended up in the lozenges, sickening over two hundred people and killing as many as thirty. This, of course, was not the first time arsenic had killed people. Members of the Borgia family, notorious Spanish aristocrats of the late Renaissance, dispatched their enemies with arsenic, and Madam Toffana of Sicily built herself a career as a poisoner in the seventeenth century. Her method was quite inventive. She rubbed arsenic into the joints of freshly slaughtered swine, removed the synovial fluid, and used it to make her Aqua Toffana. While she sold this potion as a remedy for excessive redness of the cheeks, her customers could also, if they so desired, use it to remove a spouse from the conjugal bed. Permanently. As many as six hundred people may have met their end in this fashion before the authorities brought Toffana to justice and sentenced her to public strangulation.

Arsenic really can make the skin whiter. It does this by destroying red blood cells. Victorian ladies could not abide the thought of being mistaken for sunburned peasants, so they dosed themselves with "arsenic complexion wafers." But aristocratic women were not the only ones who consumed arsenic — many other people did as well, firmly believing that it would improve their health. Stories about the healthy peasants of Styria, a region of Austria, had captivated their imaginations.

Taking arsenic on the sly, the Styrians supposedly enjoyed protection against disease, an abundance of energy, beautiful complexions, and sleek hair. Why did they have to dose themselves in secret? Because the Church considered self-medication to be a sin. Illness was due to the action of demons, and such matters had to be dealt with by the Church alone.

Many scientists thought that the practice of eating arsenic was a myth, because nobody could survive the doses that the alleged arsenic eaters claimed to be consuming. But, in 1875, a physician presented a pair of arsenic eaters at a congress of German scientists and physicians. The two men, in full sight of the gathering, proceeded to eat about four hundred milligrams of arsenic oxide each. This was at least twice the dose established as lethal. The demonstration proved that a person could build up a tolerance to arsenic by consuming small, but increasing, amounts. Dorothy Sayers, the famed British mystery writer, was captivated by the story, and she based one of her most famous works, *Strong Poison*, on an ingenious case of arsenic poisoning. The murderer shares a poisoned meal with his victim, but he survives because he has gradually immunized himself with small doses of arsenic trioxide. Lord Peter Wimsey, Sayers's detective, knows all about the nuances of arsenic and is therefore able to determine exactly how the crime was committed.

Did the Styrians really improve their health by taking arsenic? In a word, no. While small amounts of arsenic can promote growth — indeed, arsenic compounds are additives in pig and poultry feed — the amounts that the Austrian peasants ingested were dangerous. Arsenic interferes with the absorption of iodine, and it can cause a condition known as goiter, an enlargement of the thyroid gland. Many Styrians were afflicted with this condition. Even worse, cretinism, a condition characterized by stunted growth and mental retardation caused by a deficiency of thyroid hormone, afflicted many of their children.

The Styrians had easy access to arsenic because they were miners and metal smelters. Arsenic is present in many metallic ores, and when these are smelted, arsenic is converted to arsenic oxide. This substance is volatile; it emerges as a white smoke, which can be condensed on cool surfaces. Indeed, the development of the German mining industry made large amounts of cheap arsenic oxide available. You may think that this would have led to mass poisonings, but it may actually have initiated quite the opposite effect. The plague, which had devastated Europe up to the seventeenth century, came to a relatively sudden halt. This horrific disease was carried by fleas, which lived on rats. And there were rats everywhere. There was no simple way to get rid of them — at least, not until cheap arsenic oxide came along. Rapidly becoming the rat catcher's primary weapon, the poison put a huge dent in the rat population. It is interesting that one of the first industrial pollutants may have played a role in improving public health. And arsenic may still have a role to play in health improvement. Researchers at Memorial Sloan-Kettering Cancer Center in New York City report that in a trial, eleven out of twelve patients with acute promyelocytic leukemia experienced complete remission after treatment with small daily doses of arsenic trioxide.

While arsenic may have limited benefits, it is far more likely to kill than to cure. We now recognize it as a carcinogen, capable of causing skin cancer, lung cancer, and bladder cancer. This is a major worry in some areas of the world, such as Bangladesh, where the water in two to four million wells has dangerous levels of arsenic. The water consumed by up to forty million people contains a concentration of arsenic hundreds of times greater than that considered safe — about ten micrograms per liter. Ironically, authorities initiated the well boring because so many people were dying from gastrointestinal diseases brought on by drinking pond and river water contaminated by sewage.

We began to recognize the extent of the problem in 1995, and we now know that for over a decade the villagers have been drinking water containing unacceptably high levels of arsenic. Among these people, skin blemishes, lung disease, skin cancer, and liver failure are on the rise. In a few more years, we may be able to attribute one in ten deaths in this part of the world to arsenic poisoning. The arsenic occurs naturally in the soil and leaches into the well water. There is certainly no evidence that these unfortunate people are developing any sort of immunity to arsenic, but they didn't start with small doses.

Still, Dorothy Sayers's story does present us with some interesting chemistry. While Sayers wrote fiction, we all know that the truth can sometimes be stranger. Pope Clement VII was supposedly murdered in 1534 with arsenic. The killer blended arsenic oxide with the wax used to make a candle the pope would carry in a procession, and the fumes ultimately poisoned the pontiff. The perpetrator of this assassination was even cleverer than you may guess. Burning wax is a source of the hydrogen needed to change arsenic oxide to a gaseous form of arsenic, known as arsine. Arsine can kill more effectively through inhalation than arsenic can through ingestion. Arsenic is "strong poison" indeed.

A MOUTH FULL OF MERCURY

I'm not exactly a disinterested bystander when it comes to the dental amalgam controversy. My mouth is filled with "silver." So, in 1990, I was riveted to the TV when *60 Minutes*, the usually excellent news program, introduced a story with the sensational headline "Is there poison in your mouth?" It's a valid question and one that deserves scientific scrutiny.

So, is there poison in my mouth? There sure is. Dental fillings are about fifty percent mercury, a nasty metal that has the ability to wreak havoc with the nervous system. In 1953, in Minimata, Japan, thousands of people ate fish that had been tainted with mercury due to a local chemical company's reckless methods for disposing of the metal. The severity of their reactions was determined by the amount of fish they consumed: those who ate little were unaffected; those who ate a lot became totally disabled or died. In other words, mercury is only a poison at large doses. The Minimata victims stand as a tragic reminder to us all of what mercury can do.

Questions about fillings, therefore, should address the amount of mercury involved and how much of it is released into the body. Unfortunately, the questions raised in the public domain rarely do this. The *60 Minutes* episode, for example, caught the audience's attention with the account of a lady who supposedly contracted multiple sclerosis, had her amalgam fillings removed, and went dancing the next night. Nobody challenged this. Nobody mentioned that removing fillings actually releases more mercury into the system — the patient inhales mercury vapors generated by drilling.

The show's producers presented anecdotal evidence to suggest that dental amalgam can bring on depression, irritability, listlessness, and arthritis. They displayed pictures of happy individuals who claimed that by removing their fillings they had turned their lives around. But the *60 Minutes* people had not followed up on their subjects. No one had bothered to find out whether the lady had kept dancing or whether the other improvements had persisted. At least not until *Dateline*, another news show, tackled the issue in a much more scientific fashion. *Dateline*'s producers focused on the activities of a dentist named Dr. Hal Huggins, the guru of the antiamalgam movement.

Huggins claims that mercury can cause ailments ranging from prostate problems to leukemia and heart disease, and he recommends a variety of dietary supplements to rid the body of the evil toxin, an approach that has no chemical merit. "May you never know what we're preventing" is one advertising slogan.

The *Dateline* episode featured a video provided by Huggins about an MS victim who came for treatment in a wheelchair and went home with a walker. This time, the reporter did follow up, finding that as soon as the lady got home, her condition reverted; her health has deteriorated progressively since. Authors of a recent German study examined forty patients who were convinced that they had health problems due to their amalgam fillings. They compared these people with forty others who had similar fillings but no problems. The researchers observed no differences in the mercury concentrations in the saliva, blood, and urine of their subjects, but they did note differences in their psychological profiles. Many of the "victims" had obsessive attitudes towards their bodies, and they had read extensively about the dangers of amalgam.

I do not mean to say that mercury in fillings cannot cause legitimate problems. In some cases, it can. There is the classic case of the woman whose doctor diagnosed her with trigeminal neuralgia, a neurological condition that causes sporadic, intense facial pain. As it turned out, she'd had a tooth filled that was very close to a gold crown. Dissimilar metals can form an electric cell and generate current when a medium capable of conducting electricity connects them. Acidic foods can turn saliva into such a medium. Whenever the unfortunate woman ate these foods, she would suffer a jolt of pain. When her dentist replaced the amalgam with porcelain, the problem vanished. Anyone who has ever chomped on a piece of aluminum foil knows what I'm talking about.

Then, just as some people are highly sensitive to peanuts or monosodium glutamate, others react to trace amounts of mercury — amounts that do not disturb the vast majority of the public. Some sufferers of myasthenia gravis, for example — a neurological disease associated with tiredness, slurred speech, and blurred vision — have shown clinical improvement after having their amalgam fillings removed. Perhaps they have a genetic susceptibility. Unfortunately, for every such case, there are numerous others for whom the removal of fillings has no effect at all — of course, these cases don't make the news. We don't hear about the lady with Lou Gehrig's disease who spent ten thousand dollars to have her fillings extracted, only to see dental pain added to her list of problems, and we hear nothing about the MS victims who did not go dancing the day after their fillings were removed.

The New Jersey Dental School has developed an artificial mouth with artificial saliva, and it's the most accurate tool we have to measure how much mercury fillings release. Measurements show that a person must have 135 fillings to reach toxic levels; other components of amalgam combine with mercury and greatly reduce its potential to evaporate. Researchers have also carried out many studies on mercury in the urine. These studies show that the lowest levels at which we can observe mercury-related health problems is one hundred micrograms of urinary mercury per gram of urinary creatine (a common measure). Extensive amalgam in the mouth results in about four micrograms per gram of creatine in the urine. Add to this the fact that dentists have higher blood levels of mercury, live on average three years longer, and experience the same disease patterns as the rest of the population. So, I'm leaving my fillings just where they are, but I'm taking good care of my teeth to ensure that I won't need any more.

GET THE LEAD OUT

Would you believe that the Roman Empire crumbled due to a lack of sugar, and that an extreme fondness for sauerkraut brought the British Empire to its knees? I'm not sure I believe it, but I have come across a fascinating, albeit contentious, theory that supports this take on history. In those early times, most Romans had a sweet tooth, but they didn't know about sugar. So they boiled down grape juice to make a sweet syrup called "sapa," which they added to wine and to various foods. Because experience had taught them that it enhanced the sweetness of the finished product, they used lead vessels for this task. The Romans were unaware of the chemical nuances of the procedure, but we now understand that acids — such as those found in grape juice — leach lead from containers. Indeed, we sometimes refer to lead acetate as "sugar of lead."

One of lead's amazing characteristics is its extreme toxicity. Regular intake in the milligram range can have catastrophic consequences. Stomach cramps, weakness, headaches, irritability, loss of appetite, anemia, high blood pressure, and kidney problems are all classic symptoms of lead poisoning. These are nasty enough, but perhaps the most insidious aspect of lead is that it can be devastating over the long term at intakes too low to cause overt symptoms. Scientists have connected mental confusion and impaired judgment with chronic low-level exposure.

The Roman ruling classes suffered the effects more acutely because they drank more sweetened wine, usually from lead-based pewter mugs, and they had access to water delivered through lead pipes. Our word *plumbing* actually derives from the Latin *plumbum*, meaning "lead," and that is also why this element's chemical symbol is Pb. The Romans used lead to make water pipes, hence the derivation of the word. So, we can easily see why historians link Caligula's orgies, the poor military

decisions made by Roman emperors, and Nero's fiddling while Rome burned (in short, the fall of the empire) with lead exposure. While this may sound fanciful, archeological evidence does exist to bolster the lead-poisoning theory.

In part, lead is toxic due to its chemical resemblance to calcium. Calcium is essential for myriad biochemical reactions, and it's one of the main building blocks of bone. When lead is present in the system, our bodies can mistake it for calcium and incorporate it into bone, where x-rays can detect it readily. Remember how lead shields always stymied Superman's x-ray vision? Archeologists have x-rayed some ancient Roman skeletons and found abnormal levels of lead. And abnormal levels of lead may also have caused the sun to set on the British Empire. George III was king of England during the American Revolution, and some historians have suggested that it was his inept handling of the situation that led to the loss of the colonies.

George was not only inept, but he was also close to being insane. His madness has been widely attributed to porphyria, a disease of faulty hemoglobin production. It is generally an inherited condition, often characterized by purple urine — and we know that George exhibited this symptom. But lead poisoning can also cause similar symptoms. And how was George exposed to lead? Being of German ancestry, he loved sauerkraut. His cooks prepared it in lead pots — almost exclusively for the king's consumption, since sauerkraut was not a delicacy favored by the members of his royal court. It is conceivable that George alone suffered the serious effects of lead poisoning.

The question of whether lead played a part in the downfall of the Roman and British Empires may be just an academic one. But the question of whether our own lives are under attack by this stealthy intruder is one of great practical importance. Peeling paint and crumbling walls may not sound like weapons of war, but they can mount a pretty effective assault on health. Lead carbonate was the main white pigment used in paint up until about 1950, and it was still in general use until 1980, when it was finally phased out. Concern was mounting, because lead from paint was finding its way into people's bodies, particularly the bodies of young children in poor neighborhoods who had a fondness for putting flakes of peeling paint into their mouths. One milligram of lead from paint, ingested daily, can cause poisoning. Paint dust will accumulate on a windowsill as one opens and shuts the window. A child who touches this and then sucks his or her thumb can take in dangerous amounts of lead. And here is the really scary finding. The poisoning may not make itself immediately apparent. Indeed, it may show up only years later as aggressive behavior, learning disabilities, poor speech articulation, hyperactivity, or a subtle drop in IQ.

A study carried out at the Children's Medical Center in Cincinnati showed that lead can damage a young child's ability to learn and reason at exposures far less than those deemed safe by the U.S. government. There is no threshold effect for adverse reactions. Researchers tested blood lead in 276 children from infancy, and they administered an IQ test at age five. They noted a decreasing trend in the children's IQ, beginning at half the level health authorities deem acceptable. The effect may be lifelong, because early lead exposure can change the brain's hard wiring.

Millions of children in North America have high blood levels of lead, and they don't necessarily live in old, dilapidated buildings. Some parents have inadvertently made their children sick by engaging in home renovations that generate lead-laden dust. Some children have put paint flakes from old, peeling playground apparatus in their mouths. In one documented case, a family purchased a Victorian farmhouse in upper New York state and began renovating it. Soon, the family's ten-year-old dog began shaking and twisting. When the family told the vet that they had been stripping paint, he suspected lead poisoning. Blood tests confirmed that the dog and every family member had elevated levels of lead. All were treated with calcium disodium ethylene diamine acetic acid (EDTA) to eliminate the lead, but the dog died of kidney failure.

Although such drugs can rid the blood of lead, any lead deposited in the brain remains there forever, permanently impairing mental performance. And it is not only children who are affected. One of my colleagues — ironically, an analytical chemist — experienced this tragic phenomenon firsthand. When dust from renovations to the stately old residential building in which he lived drifted into the apartment he and his family inhabited, the trouble started. At first, the dust merely presented an annoying cleaning problem, but soon his wife and mother-in-law began to experience symptoms consistent with lead toxicity. The lead-filled dust became a life-changing nightmare.

We have to take lead pollution seriously. Only properly equipped and trained contractors should renovate houses that contain lead paint. A program designed to screen the blood of

children in underprivileged areas may be the best tool to identify problem homes. Massive federal funding will be needed to
remove the paint, and in some cases even the surrounding soil,
of any house that harbors lead paint chips. Experts will have to
evaluate children's nutritional status; an adequate calcium and
vitamin C intake significantly reduces lead absorption, but it is
precisely these nutrients that poor children lack. Communities
must monitor drinking water and install filters if lead levels rise
above ten parts per billion. Now that we have banned lead in
gasoline, paint, water-pipe welds, and food-can solder, we need
to study other potential sources of contamination — such as
the lead weights we use to balance car tires. The quantity of
lead that these weights shed each year can exceed fifty kilograms
per kilometer of road. As the weights wear out, they generate
lead dust that people can inhale, and people or their pets can
track the dust onto carpets where babies crawl.

Nero fiddled while Rome burned, possibly because he suffered from lead poisoning. Modern chemistry has given us ways
to monitor lead in our environment as well as techniques to deal
with the problem appropriately. Sweeping paint dust under
the carpet is tantamount to fiddling while lives — especially
children's — burn.

Lead may even have affected Beethoven, one of the world's
greatest composers. As I already noted, lead is so toxic that
even ingesting it in small amounts can cause a wide range of
symptoms. These include abdominal pain, nausea, slowed
reflexes, weakness, vertigo, tremors, loss of appetite, depression,
confusion, irritability, anxiety, and, in children, learning difficulties. And guess which symptoms Beethoven suffered from?
From the age of twenty, the composer constantly complained of
stomachaches, digestive problems, depression, and irritability.
Doctors found themselves at a loss for an explanation. But now,
we may finally have one. When Beethoven died, at the age of

fifty-six in 1827, a young music student clipped a lock of his hair as a souvenir. The student's descendants passed the memento down, until it was finally sold at auction in 1994.

The buyers decided to subject the hair to chemical analysis to see what they could discover about the great man. Did he die of syphilis, an explanation that was widely reported? If so, his doctors would have treated him with mercurial drugs, and these leave traces in the hair. The scientists who analyzed Beethoven's lock of hair found no mercury residue. Neither did they find opiate residues, demonstrating that Beethoven took no pain killers and struggled through his various ailments unassisted, fighting hard to keep his mind clear for writing music. But the scientists did report one surprising finding: there were high levels of lead in the composer's hair. They found about sixty parts per million — roughly one hundred times the amount in an average person's hair today.

How the composer slowly poisoned himself we can only guess. Perhaps he often drank from lead mugs, which were common at the time. Even today, some people may be exposed to the lead that leaches out of improperly glazed pottery or the lead-crystal decanters in which alcoholic beverages are sometimes stored. Maybe Beethoven's drinking water flowed through lead pipes. Maybe he was somehow exposed to lead-based paints. We will probably never solve this mystery, but there is a good chance that plumbing silenced one of the greatest composers in the world. And there's another Beethoven mystery that will never be solved. How did he compose his brilliant pieces given the fact that for most of his life he couldn't hear a sound? Ludwig van Beethoven, as you may have heard, was deaf.

OUR DAILY DIOXIN

Dioxin is the most toxic substance we humans have ever produced. I'm going to have some for supper today. And so will you. You can't avoid it. But it won't be your last supper — not on account of the dioxin, anyway. Still, any chemical that possesses such remarkable toxic properties and that is so pervasive and persistent in the environment merits our scrutiny. While you certainly won't succumb to any acute effects of dioxin after eating your meal tonight, the jury is out on whether you'll be having fewer meals in the long run because of the chemical. It is the issue of long-term exposure to extremely low doses of dioxin that we need to address.

The term *dioxin*, as used by toxicologists, actually refers to a number of substances that share certain molecular features. They contain chlorine atoms attached to carbon atoms that are linked to each other in structures known as aromatic rings. We consider polychlorinated biphenyls, for example — the notorious PCBS — to be dioxins when it comes to discussing toxicity. And a discussion of this brand of toxicity is fascinating. It takes us all the way from Charles Darwin's living plants to Belgium's dying chickens.

Darwin's interest in evolution inspired him to undertake a number of experiments with plant reproduction. In the course of these investigations, he discovered that growth and reproduction are governed by plant hormones, molecules that cruise through the plant delivering chemical signals. The first such hormone that scientists managed to isolate, in 1928, was auxin. A flurry of agricultural research ensued, as scientists sought to increase crop yields by boosting auxin levels in plants. As they usually do when experimentation first isolates an active and potentially useful chemical from nature, chemists went to work and synthesized a number of compounds with molecular structures similar to that of auxin. They hoped that they could improve upon nature. Two synthetic auxins, with the rather imposing names of 2,4-dichlorophenoxyacetic acid (2,4-D) and 2,4,5-trichlorophenoxyacetic acid (2,4,5-T) looked interesting. Indeed, these chemicals proved useful in promoting root growth in certain plants and in preventing strawberries from falling before they ripened. But it was in the war against weeds that the chemicals would really shine.

Both 2,4-D and 2,4,5-T caused such rapid growth in broad-leaf weeds that the weeds essentially grew themselves to death. Lawns and golf greens quickly benefited, and people frolicked in weed-free splendor. Then, during World War II, the U.S. military had an idea. Trees are broad-leaf plants. Could this chemical growth stimulation cause trees to shed their leaves, just like weeds? If so, then they could spray 2,4 D and 2,4,5-T from the air as defoliants, thereby exposing the enemy's hiding places. Experiments conducted in Florida showed that the idea was workable, but the war came to an end before the military had a chance to implement the defoliation plan.

Then came Vietnam. The jungles afforded plenty of hiding places for the Vietcong, so defoliation seemed an attractive military maneuver. Experts concocted a special blend of 2,4-D and 2,4,5-T, and the military shipped it to Vietnam in large orange-colored barrels. Eventually, they sprayed forty-two million liters of this chemical — dubbed Agent Orange — over the Vietnamese countryside in what came to be called Operation Ranch Hand. Those in command didn't worry much about exposing personnel to the chemicals, because the only health problem that had ever cropped up was a rare skin disorder known as chloracne. This had come to light in 1953, when a German physician noted several cases among workers who produced trichlorophenol, the chemical used to make 2,4,5-T. The physician examined the problem in a very thorough fashion, and he discovered that the culprit was a contaminant in the trichlorophenol, a contaminant called TCDD, a specific dioxin.

Although those in command in the U.S. Air Force were aware of this problem, they didn't think it particularly important. After all, the American servicemen involved in spraying Agent Orange were not contracting chloracne. But, as the spraying went on, so did research into the effects of 2,4,5-T. Research that, in 1969, produced a bombshell: 2,4,5-T caused birth defects in mice. Actually, 2,4,5-T itself wasn't the problem; as with chloracne, the culprit was a contaminant, dioxin. Birth defects are a far more serious problem than acne; so, all of a sudden, dioxin became a hot research topic. Scientists soon determined that they were dealing with an extremely potent substance. A millionth of a gram could kill a guinea pig. Arsenic and cyanide seemed like candy by comparison. This was the most toxic synthetic substance they had ever seen.

But then the story took a strange turn. Rabbits, mice, and monkeys proved to be two hundred times less sensitive to dioxin than guinea pigs were, and hamsters were almost immune, being two thousand times less sensitive. This was remarkable species specificity. Where did humans fit in? Well, no one has, as yet, answered that question adequately; no human has ever died of an acute dose of dioxin, even though some people have been massively exposed in chemical plant accidents. But what about long-term, low-dose exposure? This is a much bigger worry. The researchers who were conducting the animal tests learned that, despite its reduced effect on some species, dioxin is the most potent carcinogen ever tested on animals. It caused cancer in every species they tested, and it did so at a lower dose than any other substance. Furthermore, dioxin is remarkably stable and fat soluble. This suggested that it builds up in the environment, and in humans.

When the researchers released these findings, authorities took immediate action to reduce dioxin exposure. Since they believed that the only significant source was 2,4,5-T, they ensured that use of this substance was greatly curtailed. (Its confrere, 2,4-D, was never contaminated with dioxin and is still used as a weed killer.) But, much to everyone's surprise, dioxin levels in the environment did not drop. There had to be another source of these nasty chemicals. And there was. People wanted white toilet paper, white tissues, and white writing paper. Manufacturers complied by bleaching pulp with chlorine. We didn't know that this created dioxins. We were also producing an ever-increasing amount of garbage, which we burned in incinerators. Waste often contains substances — such as polyvinyl chloride (PVC), a common plastic — that furnish the chlorine that combines with combustion products to yield dioxin. Metal smelters and cement kilns also contributed to airborne dioxins. Winds carried the chemical to every corner of the world. It settled on oceans and grazing fields and was detected in the fat of fish, cattle, hogs, and poultry. And, worst of all, in the fat of humans.

Then scientists discovered that dioxins in test animals not only acted as carcinogens, but they also affected the endocrine and immune systems. So it really wasn't surprising that the Belgians panicked in 2001 when their chickens began to die due to dioxins in their food. Apparently, some unscrupulous manufacturers had mixed transformer oil with the animal fat used to make chicken feed. The authorities quickly remedied the situation, but the bad taste of dioxin remained in people's mouths.

Talking about leaving a bad taste in one's mouth, how about those comments about dioxin, attributed to a physician, that are circulating on the Internet? The physician warns the public that dioxins can leach out of the plastic containers we use in microwave ovens, and he raises the prospect of cancer. The good doctor cites the example of the foam containers that were removed from fast-food restaurants a few years ago. But he's muddled the facts.

Let me try to figure out how this story got started. Dioxins are molecules that contain chlorine, so we need a source of this element if the dioxin is to form. We also need some high-temperature incineration products of organic matter to combine with the chlorine. Indeed, if we incinerate chlorinated substances, then dioxins will form. PVC is a commonly used plastic. We use it to manufacture fabrics, tubing, and various kinds of containers. If we incinerate PVCs, then dioxins can form, but in order for this to happen, the temperature must be quite high — higher than the temperatures achieved with a microwave oven. And, furthermore, the containers we use in microwave ovens are almost always made of polyethylene or polypropylene, not PVC. These two substances do not contain chlorine and therefore cannot create dioxins. The physician's statement that foam containers were removed from fast-food restaurants because of the dioxin problem is completely false. Those containers were blown with freon, a greenhouse gas — that is why the restaurants stopped using them. Nothing to do with dioxins. In fact, the restaurants started dishing up their fast food in paper containers, and guess what? The chance of dioxin contamination with paper products is greater than with foamed polystyrene; we have to bleach paper in order to make it white, and we often do this with chlorine. I remember that a few years ago people were up in arms because traces of dioxin had been found in toilet paper. But this posed no real danger — except to toilet paper eaters. In any case, you can see how those who fail to understand the whole story can blow a kernel of scientific truth way out of proportion.

Finally, here is some good news about dioxins. Levels in the environment have been dropping. The pulp and paper industry has switched processes, and now its plants release virtually no dioxin. Regulations have forced operators to run their incinerators at high temperatures, minimizing dioxin release. Studies have revealed that chemical industry workers exposed to dioxins do have an increased incidence of cancer, but only the most heavily exposed fall into this category. Workers with lighter exposures — those who have twenty times more dioxin in their bodies than members of the general public — do not have elevated rates of cancer.

Still, we have lingering concerns. A 1976 accident in Seveso, Italy, released huge amounts of dioxin. Afterwards, the exposed men of Seveso fathered a disproportionate number of girls, a fact that suggests a hormonal effect. Without a doubt, we must keep reducing our contact with dioxins. Since about ninety-five percent of our exposure comes from animal fat, switching to a low-fat diet helps. So, while you will still be having dioxins for supper tonight, you'll consume less if you boost your vegetable intake and cut down on the meat. Bon appetit!

Don't Sweat It

I must admit I had never heard of Sammy Kershaw. But that's because as far as country and western singers go, I'm still stuck on Elvis. But I understand that Sammy is pretty hot stuff. He's also an entrepreneur of sorts. He's available not only on CDs and tapes, as one would expect, but in a more unusual format as well. Sammy, you see, also comes in a bottle. At least, his essence does. Or it used to.

Starclone was a woman's cologne that contained Sammy's underarm sweat. During his performances, the singer would wear a shirt with pads sewn into it. After the show, he'd seal the pads in plastic bags and ship them off to the University of Colorado; there, chemists would extract the essence and distill it before sending the stuff on to a perfume manufacturer, who blended it with various floral fragrances. "Sammy Sauce" — as his band members dubbed the concoction — was supposed to trip the triggers of Kershaw's fans. The promoters of the cologne made vague references to pheromones, those fascinating chemicals that can initiate mating behavior in certain species, but they didn't imply that Starclone would do anything more than allow the fans a whiff of the sweet smell of success surrounding their idol.

For most people, the sweet smell of success, as far as armpits are concerned, has more to do with eliminating odors than accenting them. And in North America this translates into an industry that generates 1.5 billion dollars annually, an industry that is dedicated to saving us from the offensive vapors of others. Selling antiperspirants and deodorants is a huge business, undoubtedly because some of the chemicals these products eliminate are truly disturbing. Such as 3-methyl-2-hexenoic acid — referred to in the trade as "armpit in a jar." One can actually find that jar at the Monell Chemical Senses Center in Philadelphia, where chemists have isolated the compound from male underarm sweat and determined that it is the main odiferous component. That may well be, but it certainly isn't the only one. The chemists have identified over two hundred compounds in sweat, some of them rivaling 3-methyl-2-hexenoic acid in terms of stench potential. We could refer to a vial of isovaleric acid as "locker room concentrate," and androstenol recalls the distinct aroma of a public urinal. A lucky three percent of the population is spared such misery; these people cannot smell isovaleric acid at all. Then there is 4-ethyloctanoic acid, which reminds anyone who has ever smelled one of a wet male goat. This foul-smelling compound is a real turnoff — except, apparently, to mature female goats in heat. Catullus, the Roman poet, noted the goatish smell as early as 50 B.C., when he wrote about certain people "keeping a fierce goat under the arms." But we don't know whether Catullus was aware of the relationship between the smell and the mating habits of the animal in question.

Where do these underarm odors come from? We can blame bacteria that inhabit the surface of our skin. There are millions of them, and they feed on us. Indirectly. Our apocrine sweat glands, which are mostly found in our armpits and private regions, produce a yellowish fluid that harbors fats, proteins, and various steroids. The fluid has no smell, but its components are great food for bacteria. As they digest these components, the bacteria produce a variety of malodorous compounds. To put it bluntly, unless we take care, we'll end up reeking of bacterial poop. The apocrine glands do not become active until puberty, feeding the theory that the compounds they excrete behave as human pheromones, or sex attractants.

While the idea that some sweat glands release chemical messengers is debatable, it is clear that not all sweat glands have this capability. Most of the roughly 2.5 billion sweat glands that riddle our bodies produce a watery fluid that contains dissolved ions — such as sodium, potassium, and chloride — but no organic substances. These eccrine glands serve a single purpose, and that is to cool the body. They are active from birth, and they release moisture in response to internal heat caused by high external temperature, muscle activity, or overstimulated nerves. Changing a liquid into a vapor requires heat; when heat is drawn from the body, the body cools down. An undesirable feature of the eccrine sweat process is that along with moisture we also lose minerals. Well, we don't exactly lose them, but they are removed from our circulation. The minerals, mostly salt, remain on the skin after the water evaporates. That's why skin tastes salty after heavy exercise. If we sweat extensively, then we may need to replace minerals by consuming a sports drink.

Excessive activity in the nervous system can also cause sweating. Anxiety or fear can prompt the adrenal glands to churn out adrenaline, the "fight or flight" hormone. The nerve cells that make up the part of our autonomic nervous system that relies on adrenaline for its activity (the sympathetic system) are connected to sweat glands; this way, if they get overheated through frantic activity, then they will be cooled. This keeps them functional. Drugs that stimulate the activity of the sympathetic system can also cause sweating. In fact, that was the clue Italian police needed to solve a bizarre attempted-murder case. Doctors had been unable to find a reason for a thirty-seven-year-old man's sporadic but intense episodes of sweating and malaise. They suspected a tumor of the adrenal gland, but the solution turned out to be more sinister. Upon reflection, the patient himself realized that the episodes usually occurred after his wife, who was a nurse, brought him mineral water at lunchtime. As it turned out, she had doctored the drink with metaraminol, an adrenaline-like drug used to treat low blood pressure. There was another man involved, but he will have to do without his lover for a very long time. She was convicted of attempted murder. I bet that she broke out in a cold sweat when the police caught her.

But what about all those innocent people who have to fight underarm battles? That's where deodorants and antiperspirants come in. The two products are not identical. Deodorants contain fragrances that mask the sweat smell as well as antibacterial agents that control the growth of bacteria on the skin. Antiperspirants, however, contain aluminum compounds that form insoluble gels on the skin and plug up pores, reducing the amount of sweat that makes it to the surface. In spite of the wild allegations that some have made about aluminum-based antiperspirants on the Internet, these products have a history of safe use. Claims that they can cause breast cancer by allowing toxins to be deposited in the lymph nodes instead of being expelled through sweat are nonsensical. The body eliminates toxins via the liver and kidneys, not through sweat. In any case, the toxins that are not eliminated end up circulating throughout the body; they are not somehow deposited in lymph glands near the breast. The only link between aluminum compounds and breast cancer has to do with mammograms: the technician conducting the test may confuse antiperspirant residue with calcium deposits, but these deposits are not cause for concern.

Some unfortunate people, by their own account, sweat like pigs, no matter what they do. The expression is actually inappropriate, because pigs have no perspiration apparatus. That's why they wallow in mud. This is not a useful option for humans. Surgery, however, is. The procedure known as endoscopic transthoracic sympathectomy stops signals from being transmitted along the nerves that stimulate sweat. But doctors only resort to this in the most extreme cases. What can a person do if deodorants and antiperspirants cannot turn off that chemical tap in the armpit? Maybe consider going into business. Like Sammy Kershaw did. Starclone is no longer on the market, so there is a need for an improved version of the bottled armpit. Maybe Sammy just didn't have the right chemistry. Where is Elvis when you need him?

EVERYDAY SCIENCE

THE STAINMASTER

I was a little taken aback by the call. "Is that the Stainmaster?" the lady asked. While I'd been called names before, I had never been addressed in such an exalted fashion. I suppose I have developed somewhat of a reputation for helping people solve their stain problems through the appropriate use of chemistry, but "Stainmaster" was far too complimentary a title. It also intimidated me, because I knew I would be hearing about a particularly tough problem. I was right.

Tragedy had befallen a Barbie doll, the caller told me. She was a collector's item, and she had been purchased for a tidy sum. But there was a slight problem: her red lip had been chipped. Barbie's new owner tried to remedy the situation with a red marker, which, unfortunately, had bled into the surrounding flesh-colored plastic, making the poor creature look as if she had developed some strange allergic rash. Now this, the lady thought, was a job for the Stainmaster.

Stain removal is a fascinating chemical challenge. It is also incredibly complex. I wish I could outline a direct, step-by-step approach that one could take to remove any stain. But this is just not possible. There are simply too many variables. Different fabrics require different treatments. We cannot tackle candle wax on a tablecloth the same way we handle coffee on a shirt. And we even have to approach a coffee stain differently depending on whether milk or sugar is involved. But over the years I have amassed some experience in this area, and I am happy to share it here because it gives us a chance to explore a few fascinating facets of chemistry.

The practice of stain removal is based on four basic principles: absorbency, dissolution, detergency, and chemical reactivity. Absorbents — such as cornstarch, talcum powder, or salt — not only absorb excess liquids but also help lift fresh stains — such as red wine — from a surface. Different substances react in various ways to solvents. We can remove apple juice with hot water, but we require alcohol to dissolve lipstick dye. Greasy stains will not dissolve in pure water, but they will often dissolve if we add soap or detergent to the water. Such additives change the surface tension of the water, allowing it to flow more easily into the crevices of the fabric. At the same time, the soap or detergent molecules forge a link between the oil and the water; one end of the soap or detergent molecule binds to the oil, the other to the water. When we rinse the fabric, the stain lifts from its surface. Some stains cannot be dissolved, but they can be decolorized through appropriate chemical reactions. We can employ bleach, for example, to strip electrons from molecules. Electrons are more than just the glue that holds molecules together — they are also responsible for color. By stripping molecules of electrons, or "oxidizing" them, we can therefore eliminate stains. Another handy chemical reaction makes use of the enzymes that break down the proteins that cause some stain problems. To remove stains effectively, we may need to combine these methods.

Without a doubt, the key to stain-removal success is speed. You can get most stains out with water if you act quickly enough. If flushing with water (club soda is especially good, because the bubbling action helps dislodge the stain) doesn't do the job, rub a little Ivory soap or Ivory dishwashing detergent into the stain and then rinse. If this doesn't work, or if you are dealing with a stain that has dried, then you'll have to take the solvent step. My first choice is usually dry-cleaning solvent, or perchloroethylene, also called "perk."

At one time, I used a product that contained perk, and I just loved it. It was called K2R, and it came in an aerosol can; you sprayed it directly onto the fabric. At first, the results were frightening. The fabric turned white. But the white stuff was a powder that absorbed the solvent after it had dissolved the stain. Once it dried, you just removed it with the brush that was conveniently attached to the cap of the container. Alas, I can't find this product anymore, probably because the sale of chlorinated solvents is restricted — these substances can be toxic.

You can also try regular dry-cleaning fluid (if you can find it) or lighter fluid. Particularly effective is Goof Off, which is a mixture of xylene and various other solvents. It's great for dried latex paint, as well as many inks and glues. Place the fabric, stain side down, on a white cotton towel and drip the solvent through. If this doesn't do the job, pour some of the solvent onto a piece of white cotton and dab at the stain. Never rub! The reason I try these solvents first is because they cannot do any harm; if your stain problem persists, you'll have to battle it with other solvents. I usually try a citrus oil product like Orange TKO next. It can be a knockout. Made from orange peel, its main ingredient is limonene, an excellent solvent for greasy materials, many inks, and even candle wax and gum. Laundry Miracle is another of my favorites. Amyl acetate, also called banana oil, is the active ingredient here — one whiff and you'll know where its alternate name comes from.

The alkoxylated alcohols form another superb family of solvents. I love these too. They dissolve a wide range of stains. Some of my greatest successes have come with Spot Shot. Its manufacturers market it for carpets, but it works on other fabrics as well. It is a combo of 2-butoxyethanol and a detergent. Spray it on, wait a bit, and dab with a paper towel. Shout, in its various formulations, is also worth shouting about. I've had good luck with the aerosol, the liquid, the gel, and, especially, the laundry stick. All of these contain some version of the alkoxylated alcohols, but the stick also has some propylene glycol — another good solvent — as well as a detergent and an enzyme that breaks down protein stains. Apply it, wait a moment, and then launder.

There are numerous other enzyme-containing products on the market, such as Bio-Ad and Amaze. Make a paste with one of these, rub it into a stain (as long as it's not on wool or silk), and let it sit for a while in a warm environment. Rinse and be amazed! On occasion, you may have to soak a stained item in a solution of Amaze. Should that fail, however, bring on the bleach. My first choice is a paste made of sodium perborate (Bleach for Unbleachables); if this doesn't work, then a dilute solution of sodium hypochlorite (white fabrics only) may do the trick. Oxy-Clean, which releases oxygen when it dissolves in water, can also be effective. If all of these fall short, then get out the scissors.

All right — let's put some of what we've learned into practice. You've been drinking grape juice because you've heard that it contains resveratrol, which may protect you from a heart attack. But you've spilled some of the deep-purple liquid on your favorite shirt, just about giving yourself one. Fret not. Use Spot Shot or Shout, then, if needed, a paste of sodium perborate. The lycopene in the tomato sauce you've been eating to lower your cancer risk has left its mark. Spot Shot, then dilute vinegar, then dry-cleaning solvent will make life worth living again. You just made a diving catch in the outfield, and now you have grass stains as a souvenir. Amaze will take care of it. Ballpoint ink? Try limonene or amyl acetate. Cranberry sauce on the tablecloth? Hold the stain over a bowl and pour hot water through it from a height of about a foot. Do this carefully. Then use Shout. Does mustard on your best tie make you want to slash your wrists? Try Spot Shot, then rub with pure soap, and finish up with all-fabric bleach. If this doesn't work, and you do end up slashing your wrists, rinse your shirt in cold water, rub the bloodstains with Bio-Ad, wait fifteen minutes, and rinse with dilute ammonia.

It's a complicated business, isn't it? Especially if a stain has soaked into the material and set. That's exactly what happened to poor Barbie. I tried everything. After all, my reputation was at stake. I dabbed with hydrogen peroxide. I patted with perchloroethylene. I washed her tainted "skin" with limonene, acetone, and amyl acetate. She smelled like a fruit salad, and she still looked like a clown. I was out of weapons. The Stainmaster had failed.

But hang on. There may be hope. I've recently come across Ossengal, an "odorless, organic, chemical-free" spot remover made from the purified gall of oxen. It's hard to see how this mix of acids, phospholipids, and enzymes could be "chemical-free," but never mind. If you eat like an ox, you probably need some powerful stuff to break down your food. According to the Ossengal ads, at least, spots that are treated with the product will disappear instantly. Maybe I'll try it on Barbie. Maybe Ossengal will remove the stain on the Stainmaster's reputation.

STRIKE ONE FOR MATCHES

Picture a castle in Old England. Its dingy halls are lit by torches, and a fire struggles to warm the cold, damp air. Now give a thought to how the servant responsible for such things started that fire. It wasn't easy. Carefully, he made a little pile of tinder, probably of wood shavings or linen threads. Then he struck a flint repeatedly against a piece of metal, hoping that the sparks he produced would ignite the flammable material. Today, we would just strike a match. Oh, how we take our modern conveniences for granted! Reliable matches, those little chemical-coated sticks that have had such an impact on history, are a recent invention. Let's return, however, to our dimly lit seventeenth-century castle, the domain of King Charles II.

King Charles heard tell of a brilliant German entertainer who had a reputation for lighting up a room with his antics — literally. So he invited Daniel Kraft to amaze the royal court. And that he did. Kraft took a little bit of a soft, waxy substance from a bottle and placed it on a pile of gunpowder. Within a few seconds, as if by magic, the gunpowder burst into flame. Charles and his guests, including Robert Boyle, the famed "natural philosopher," had just witnessed a demonstration of the wondrous properties of phosphorus. (For more on this see p. 237.)

Kraft had learned the secret of concocting this amazing substance from the alchemist Hennig Brandt, who discovered it, in 1669, by allowing a vat of urine to boil dry. He had been trying to find the secret of life, which he thought was concealed in this bodily secretion. Boyle was captivated by the eerie glow of phosphorous and by its ability to burst into flame. But the only clue about its derivation that he could squeeze out of Kraft was that it came from "somewhat that belonged to the body of man." One of Boyle's apprentices remembered seeing a similar demonstration in Hamburg, and, at his master's request, he tracked Brandt down and learned the secret. Boyle immediately began to experiment with phosphorus, and he even thought of making a luminous watch dial with it. But his most memorable experiment involved dipping a wooden splint in sulfur and then pulling it through a piece of paper to which he'd applied a little phosphorus. The friction generated heat, which ignited the phosphorus, which ignited the sulfur, which ignited the wood. The match was born.

Boyle's match, however, turned out to be no more than a curiosity, because phosphorus was very hard to come by, and its ignition was hard to control. Johan Gahn of Sweden solved the availability problem when he discovered that bones are made of calcium phosphate and that by heating these in the presence of carbon he could isolate phosphorus. Soon, people could purchase little bottles coated on the inside with phosphorus. To produce a flame, one had to scrape a splint dipped in sulfur against the inside of the bottle and then rub it on a cork. Sometimes, though, all of the phosphorus, and the person who was scraping at it, would ignite. Peyla, an Italian chemist, got around this difficulty in an ingenious way. He sealed a candle with a piece of phosphorous attached to its wick inside a glass tube. He dipped the tube into hot water, melting the phosphorus, which then impregnated the wick. When he broke the glass, the phosphorus was exposed to oxygen, and the wick ignited. While these inventions were clever, a serendipitous discovery made in 1825 by an English pharmacist was to eclipse them all.

A customer had asked John Walker to make up a mixture of antimony sulfide, potassium chlorate, and vegetable gum because he had heard that this mix would ignite when struck. Walker did as requested, and he noted that a tear-shaped drop formed on his stirring rod. When he tried to scrape it off, it burst into flame. Walker began to make matches based on this method, but he never patented the process. Samuel Jones saw a demonstration of Walker's invention and managed not only to capitalize on it but to improve on it as well. He added sulfur to the mix, making his "lucifers" easier to ignite — but at a cost. The user had to put up with the choking stench of sulfur dioxide produced by the burning sulfur.

Then, Charles Sauria of France had an idea. By adding a little easily ignitable phosphorus to the mix, he could reduce some of the offensive components. Soon, these newfangled lucifers flooded the market. They also caused hospitals to be flooded with miserable patients. White phosphorus is highly toxic, and people who made matches routinely developed "phossy jaw," a terrible condition in which the jaw bone disintegrates. Then further experimentation yielded the discovery that white phosphorus, when heated in the absence of air, converts to red phosphorus. The red variety is not toxic, and it ignites at a much higher temperature. It didn't take long for this observation to trigger the formulation of safety matches. Manufacturers would dip a splint in paraffin, then coat it with a mix of sulfur (a good fuel), potassium chlorate (to provide oxygen), and glue. One ignited the match by rubbing it on a strip of red phosphorus and ground glass (for friction) that was attached to the outside of the matchbox.

And so it was that by the end of the nineteenth century, the flint and the tinderbox were relegated to the dustbin of history. Matches had struck a chord with everyone. And, with the wide availability of safety matches, people no longer had to worry about one of their lucifers igniting in the cupboard when a rat gnawed on its tip. Not unless the rat knew how to remove a match from the box and strike it against the phosphorus-coated strip.

COOKWARE CHEMISTRY

Like most chemists, I like to cook. After all, what is cooking but the appropriate mixing of chemicals? In the lab, we use flasks and beakers, but how do we equip our kitchens? At Tiffany's in New York City, you can procure a silver frying pan for thousands of dollars; specialty stores sell gleaming copper pots for a couple of hundred, while you can buy a thin aluminum pot most anywhere for a few bucks. What's the difference?

In a trembling voice, the gourmet cook will describe how hot spots on the bottom of a pan can destroy a delicate sauce. But we can avoid such tragic outcomes by choosing cookware that allows for efficient heat conduction and precise heat control. Silver is an ideal material for this, but it's very expensive. Copper conducts heat almost as well, but it can dissolve in food and cause nausea, vomiting, and diarrhea. Copper cookware lined with tin or stainless steel is the solution; however, this lining eventually wears off and has to be replaced. Cookware manufacturers also make aluminum or steel pans with a thin layer of copper on the bottom, but such a small amount of copper does not improve the pan's heat conduction.

Aluminum itself is an excellent conductor. In fact, more aluminum cookware is sold than any other type. But some cooks have discarded their aluminum pots and pans in response to reports linking aluminum with Alzheimer's disease. Most researchers today do not believe that aluminum is a causative factor in Alzheimer's, and, in any case, tossing out our aluminum cookware would have no effect. Most of it has a nonstick finish, but even people who use uncoated aluminum for both cooking and food storage ingest only about 3.5 milligrams of the metal every day. Compare this with the roughly twenty milligrams we consume daily as a natural component of food, or to the one thousand milligrams found in a daily dose of the antacid we take for an upset stomach. A single buffered pain reliever contains more than ten milligrams of aluminum.

If you want to put your mind completely at ease, then you should avoid cooking or storing highly acidic or salty foods — say, tomato sauce, rhubarb, or sauerkraut — in uncoated aluminum pots for long periods. In fact, a good general rule is to store food in the refrigerator in glass containers, not in pots or pans. Even though storing foods in metal containers poses no danger, aluminum pots will sometimes discolor as they react with food. To restore a discolored pot's finish, try using it to simmer a couple of spoonfuls of cream of tartar dissolved in a quart of water for fifteen minutes.

A new kind of aluminum cookware, made of anodized aluminum, has recently arrived in the marketplace. It is usually gray-colored, it's harder than stainless steel, and it conducts heat better; it's also eternally nonstick, scratch resistant, and easy to clean. The process of anodizing involves passing the aluminum through a series of electrochemical baths, which cause a hard layer of aluminum oxide to form on the surface. This layer is nonreactive, and it does not leach aluminum into food.

The original nonstick coating was Teflon, discovered accidentally in 1938. This inert substance does not react with food in any fashion — words of comfort to anyone who has ever worried about swallowing the bits of Teflon coating that flake off older cookware. The perfluorocarbon resin passes through the body unchanged. Teflon may, however, release some toxic vapors if heated to high temperatures for long periods. While this represents little risk to humans, there have been isolated reports of pet birds being overcome by the fumes.

Just as Teflon cookware represents modern times, so cast iron cookware symbolizes traditional cooking. Cast iron is a good conductor of heat, and, since small amounts of it dissolve into food, it even serves as a dietary source of iron — one of the few nutrients in short supply in the North American diet. We should season our cast iron cookware to keep it from rusting and to prevent food from sticking to it. To do this, coat the pan with a thin layer of oil and then heat it. The oil will react with oxygen to form a tough, smooth, impervious layer.

Stainless steel is durable and does not tarnish. We make it by alloying iron with other metals, most notably nickel and chromium. Since stainless steel's heat conduction tends to be uneven, most pieces of stainless steel cookware have an aluminum or copper bottom. Some nickel may leach into acidic foods like applesauce, and this presents a problem, especially to people with nickel allergies. We don't need to worry about chromium, though, because most of us could use a little more of this mineral in our diets.

To minimize leaching, manufacturers sometimes coat steel with porcelain — thus creating enameled cookware. The finish is stain and scratch resistant, and it's perfectly safe. Manufacturers use no lead in such glazes; while they may use lead in glazes destined for slow-cooking equipment — such as crock pots — they must guarantee that no leaching of the metal will occur. The scary stories we sometimes hear about lead poisoning from ceramic cookware invariably involve improperly manufactured items.

One classic story involves members of a California family who suffered acute lead poisoning after drinking orange juice stored in a ceramic pitcher purchased in Mexico. At first, no one was able to diagnose the condition. Their doctors advised them to leave their house, just in case the problem was environmental. When the family complied, the symptoms disappeared. When they returned to the house, the symptoms reappeared. This cycle occurred several times. Then the family called in a chemist, who quickly realized that the source of the problem was the lead glaze on the juice pitcher. Obviously, they should have consulted a chemist sooner.

So what is the best cookware? The answer depends on one's personal preference and one's pocketbook. Trained French chefs worship copper. But for ordinary kitchen chemists like me, a stainless steel pan with a thick aluminum bottom is just fine. Anodized aluminum is also excellent. Nothing sticks to it, it cannot be scratched, and it's a snap to clean. And, as far as concerns about Alzheimer's disease go, I've been cooking with aluminum pots all my life, and I can't remember experiencing any problems.

NERVOUS ABOUT NERVE GAS

I am nervous about nerve gas. It is a terrible weapon, and it's not beyond the grasp of terrorists. In 1939, a German chemist named Gerhard Schrader discovered the first nerve gas. Searching for better methods to control insects, he chanced upon a substance that had greater insecticidal activity than anything he had ever seen. He named the new compound "tabun" and envisioned a breakthrough for agriculture. Hitler, however, had something else in mind for the substance. If it could kill pests, it could also kill people. A terrible new weapon was born.

Tabun is a colorless, odorless, relatively volatile liquid. Exposure to a few milligrams is enough to cause death. It penetrates intact skin without any irritating effect, so that a person can unknowingly absorb a fatal dose. The term "nerve gas" was used to describe the substance because of its mechanism of action. Tabun interferes with the way information is transmitted from one nerve cell to another. Such transmission involves the release of chemicals called neurotransmitters from nerve endings; these then migrate across the tiny gap that separates nerve cells, known as the synapse. A neurotransmitter stimulates an adjacent cell by fitting into a receptor site on its surface, very much the way a key fits into a lock. This cell then releases a neurotransmitter, which stimulates the next cell, and thus the message is propagated. The specific neurotransmitter involved in the nerve gas story is acetylcholine.

Once acetylcholine has carried out its job of triggering a reaction in an adjacent cell, an enzyme present in the synapse decomposes it. Overstimulation is therefore prevented. It is this enzyme, acetylcholinesterase, that nerve gas deactivates. The result is overstimulation of the nervous system, eventually leading to convulsions, paralysis, and respiratory failure. The first symptoms of exposure generally include constriction of the pupils, dimming of vision, vomiting, sweating, defecation, the release of secretions from the nose, eyes, mouth, and lungs, and muscle twitching. Inhalation of the gas causes death within minutes, but the effects of liquid exposure may be delayed as much as eighteen hours.

By the end of World War II, the Germans had developed sarin, a nerve gas far more potent than tabun. The chemistry was relatively simple: methylphosphonyl difluoride mixed with rubbing alcohol. But how could they safely store such a ferociously toxic substance? With the introduction of binary weapons after the war, the German military devised an ingenious solution to this problem. The actual mixing of the gas would take place inside a missile or artillery shell after launch. The two chemicals would reside in compartments separated by a barrier designed to rupture upon acceleration, after the projectile had been fired. During flight, the shell or missile would rotate at fifteen thousand revolutions per minute; the components would blend together, forming sarin.

Unfortunately, terrorist groups can obtain the chemicals required to make sarin with relative ease. They can synthesize the key component, methylphosphonyl difluoride, from dimethyl methylphosphonate, which is commercially available since it is widely used to make flame retardants. These terrorists may not have the technology to design a sophisticated binary weapon, but this is not really necessary. In 1995, terrorists pulled off an attack in the Tokyo subway using a primitive system that involved puncturing two plastic bags containing the required chemicals and mixing them. Luckily, the mixing was not very effective, or the death toll would have been much higher than twelve. But even this crude delivery system injured over five thousand people.

The main defense against nerve gas is protective apparel. Gas masks with charcoal filters can reduce the concentration of the gas in inhaled air by a factor of about 100,000. Specially made military protective clothing impregnated with charcoal is also available. The surface of the fabric is treated with a wetting agent that causes droplets to spread out, enhancing evaporation. Since nerve gases break down rapidly in alkaline solution, decontamination of exposed surfaces with hypochlorite (bleach) or bicarbonate (baking soda) is at least a theoretical possibility. The military distributed decontaminating powders incorporating such chemicals to populations at risk of attack during the Gulf War.

Scientists have investigated antidotes for nerve gas poisoning extensively. Since the 1930s, the first line of defense after exposure has been the atropine injection. This compound is an acetylcholine antagonist, because it dislodges acetylcholine from receptor sites, reducing the risk of overstimulation. U.S. military personnel during the Gulf War carried three automatic injectors loaded with two milligrams of atropine each; several doses may be required to reduce the severity of the symptoms caused by exposure.

An atropine injection alone is effective for only a short time. Since the nerve gas deactivates acetylcholinesterase, the concentration of acetylcholine will keep increasing, and it will eventually overpower the protective effect of atropine. One must administer a second substance, pralidoxime chloride, to release the nerve gas from the enzyme and destroy it. This, too, is available in automatic injectors, and the Gulf War soldiers carried three of them for concurrent use with atropine.

Even those who survive nerve gas exposure may suffer convulsions, and for this reason an automatic injector containing ten milligrams of diazepam (Valium) is generally used with the third dose of atropine. Unfortunately, the timing of these injections is critical. By the time a victim recognizes the signs of nerve gas intoxication, it may already be too late to take the antidote. Even survivors may not get off scot-free: researchers working with animals report some evidence that sarin can cause cancer.

While, in theory, the antidotes should work, no one has actually ever tried them under battle conditions. But, sometime, somewhere, the test will come, because, as we have seen, sarin is not all that hard to synthesize. And we know what terrorists are capable of.

How Many People Does It Take to Invent a Lightbulb?

Ask who invented the lightbulb, and most people will mutter something about Thomas Edison. But Edison did no such thing. Scientists were tinkering with lightbulbs for decades before Edison came along. America's greatest inventor did, however, develop the first lightbulbs suitable for widespread use and the electrical grid needed to power them.

Way back in 1802, Humphry Davy, one of the most brilliant chemists of all time, became interested in the phenomenon of electricity. By this time, he had already published a treatise on the pain-killing properties of laughing gas and suggested its use in surgical operations. As a result of his chemical prowess, the Royal Institution in London invited Davy to further his career there. The institution's mandate was to encourage scholars to combine "useful knowledge with the amusement and instruction of the higher ranks."

Davy had closely followed the work of Italy's Alessandro Volta, who had just discovered that if he grasped wires attached to silver and copper plates separated by moist cardboard he would experience a shock akin to that caused by grabbing an electric eel. Volta had created a primitive battery. Davy recognized this effect as an excellent device for combining useful knowledge with amusement. In the basement of the Royal Institution, he installed a battery made of two thousand of Volta's "cells" and noted that a wire connecting the ends began to glow red. Here was a method to produce both heat and light. Could it signal the end of the inefficient and often dangerous gaslight?

Unfortunately, the glow did not last long because the wire melted. No material available to Davy could withstand the intense heat that the device produced. But Davy did observe that sometimes, as the wire melted through, a spark would jump between the two disconnected ends. Intrigued by this phenomenon, he experimented further and discovered that if he interrupted the wire with two pieces of charcoal placed at an exact distance from one another, he could generate a continuous spark, or an electric arc. The light produced by the sparks and the glowing charcoal was brilliant.

Before people could put arc lamps to practical use, however, Davy had to make yet another discovery — namely, Michael

Faraday. As a youngster, Faraday worked for a bookbinder. One day, a customer brought in a volume of the *Encyclopedia Britannica* for repair, and, as luck would have it, the book fell open at the entry on electricity. Faraday was captivated and wanted to learn more. He began to attend Davy's famous public lectures at the Royal Institution, and the great man soon hired him as a lab assistant. Faraday went on to enjoy a brilliant career, unraveling many of the secrets of electricity and magnetism. Perhaps his greatest discovery was that by moving a coiled wire in a magnetic field, he could cause an electric current to flow through the wire. In short, he'd invented the electrical generator, or dynamo.

Arc lights could now proceed full steam ahead. Steam-powered generators delivered the needed electricity, and soon London streets basked in an almost daylight glow. But even though the brilliant light dazzled the public, many scientists had long understood that arc lights were impractical; they needed constant servicing and were much too bright for home use. These scientists had not forgotten Davy's observation of the glowing wire. Could they find a substance that would not melt when a strong current flowed through it?

As early as 1841, Englishman Frederick DeMoleyns patented an electric bulb that featured powdered charcoal between platinum wires. DeMoleyns's bulb produced light for a few minutes, but then the charcoal burned. His countryman Joseph Swan improved on it in 1848 by fabricating filaments from paper strips coated with charcoal, which he baked at a high temperature in a pottery kiln. And then Swan had a bright idea. If he pumped the air out of the bulb, then there would be no oxygen present, and the charcoal could not burn. Since vacuum pumps had recently been invented, he was able to test his idea. In 1878, Swan demonstrated an evacuated bulb with a glowing carbon filament.

And this is where Edison enters the picture. Hundreds of inventions, including the stock ticker and the phonograph, had already made him rich and famous. He now began to devote his energy and his massive resources to taking a scientific curiosity — namely, the lightbulb — and turning it into a practical commodity. Edison, by sheer force of will, accomplished this. His "insomnia squad" of workers tried over sixteen hundred different filaments, hoping to find one that could stand up to the electric current. These ranged from beard hair to carbonized sewing thread. Finally, in 1879, just fourteen months after beginning his search, Edison stripped a thin piece of bamboo from a fan and heated it in an oxygen-free environment until it carbonized. The bamboo filament glowed in an evacuated bulb for forty hours.

Edison's genius, however, lay not only in producing a practical bulb but also in rapidly envisioning and designing a complete electrical distribution network, from the power plant and the transmission wires to home switches and sockets. On New Year's Eve 1880, he dazzled more than three thousand onlookers when he lit up his Menlo Park, New Jersey, "invention factory" with forty bulbs. Within two years, Edison had built the Pearl Street Generating Plant in Manhattan and was supplying electricity to eighty-five houses, shops, and offices. As he worked, Edison had carefully patented his progressive discoveries; Swan had not. Edison now had the temerity to sue Swan for patent infringement, in spite of the fact that Swan had built his first lightbulb when Edison was only one year old. Eventually, the two settled their differences and collaborated on the commercial production of "Ediswan" bulbs.

Over time, the Ediswan bulb has been improved upon. Filaments made of tungsten, the metal with the highest melting temperature, were introduced in 1911. Tungsten bulbs burned for hundreds of hours before their filaments evaporated, often producing the characteristic black deposit on the inside surface. Filling the bulb with a mixture of argon and nitrogen slowed this evaporation and allowed for a product that could function at higher temperatures and give off more light. Then, if manufacturers added a halogen, such as iodine, to the bulb, it reacted with the evaporating tungsten atoms and redeposited these on the filament. Halogen lamps could therefore operate at even higher temperatures and produce even brighter light. But the bulbs had to be made of more expensive quartz, because ordinary glass would melt at these temperatures.

Today, tungsten-filament lightbulbs routinely last for 750 hours. But in the offing is a bulb that has no filament and burns for over ten thousand hours. The bulb is made of quartz and filled with argon and a small amount of sulfur. The sulfur atoms produce a stunning amount of light when excited by microwaves. Prototypes already light the Smithsonian Air and Space Museum in Washington, but so far the bulbs are too bright for home use. Shades of the arc lamp.

A Run on Stockings

The moans and groans coming from the Philadelphia hotel room alarmed other guests passing by in the hallway. But by the time the manager appeared and opened the door, the sounds had ceased. Wallace Carothers, the man whose invention was the very embodiment of DuPont's famous slogan "Better Things for Better Living . . . Through Chemistry," lay sprawled on the floor, dead.

One of the most popular attractions at the 1939 New York World's Fair was the DuPont exhibit. There, steel balls came crashing down onto sheets of glass laminated with polyvinylbutyryl, a brand-new plastic. Miraculously, the glass did not shatter. This was obviously the perfect material for automobile windshields. Then there was Princess Plastic, a synthetic lady dressed from heel to hat in materials created in the DuPont laboratories. But for many a gentleman, the greatest attraction was Miss Chemistry, a shapely young lady who reclined on a couch with her skirt hitched high to demonstrate DuPont's newest invention. Ladies also flocked to this exhibit, drawn not by Miss Chemistry's legs, but by what those legs were sporting. They were absolutely amazed by the newfangled stockings, which seemed as sheer as silk but were made of a fiber stronger than steel. A fiber that — according to the company's advertising — was made of coal, air, and water. This was the triumphant debut of nylon.

Carothers never dreamed that his invention would capture the public's fancy in this fashion. He never imagined that on N-Day, May 15, 1940, women would line up at the doors of department stores across the country, eager to be among the first to purchase a single pair (one per customer) of nylon stockings. Five million ladies went home happy that day. But this was only the beginning. Soon, nylon found its way into parachutes — a critical development, since Japan had cut the U.S. off from its Oriental silk suppliers. Indeed, the military encouraged women to turn in their old nylon stockings to be recycled into parachutes. Actress Betty Grable championed this cause and auctioned off a pair of nylons that had sheathed her famous legs, fetching forty thousand dollars' worth of war bonds. The new material was also used to make ropes, tires, tents, and numerous other items essential to the war effort. Curiously, this most practical of discoveries had been the handiwork of Carothers, a man who had accepted a job at DuPont under the condition that the company would not require him to carry out research aimed at turning a profit.

Wallace Carothers had earned a doctorate in chemistry from

the University of Illinois, working under Roger Adams, probably the best-known American chemist of the era. He went on to teach at Illinois for a year before accepting a professorship at Harvard. It was the research he conducted there, into the fledgling field of polymer chemistry, that roused the interest of DuPont. Hermann Staudinger, in Germany, had made the controversial suggestion that small molecules could be linked together into long chains, called polymers, which had decidedly different physical properties from the starting material. Yet no one knew exactly what sort of forces held the small molecules together, so Carothers applied himself to solving the problem. He quickly concluded that there was no great mystery. Scientists already understood that atoms in molecules were held together by the sharing of electrons. Such covalent bonds could also be forged, Carothers surmised, between atoms of different molecules, creating a long chain.

DuPont had made its fortune producing gunpowder and explosives, but it had begun to branch into other areas. Plastics, which were made of polymers, seemed to be the wave of the future. And Wallace Carothers was one of the brightest lights in this field of research. The only problem was that he had no interest in practicalities — he admired the pure academic quest for knowledge. On numerous occasions, DuPont representatives had tried unsuccessfully to entice Carothers away from Harvard. Then they sweetened the pot. They promised Carothers that he would have unlimited research funds and as many assistants as he needed; there would be absolutely no company interference in his work; he would be allowed to pursue his academic interests; and if there was a practical spin-off, that would be gravy. Carothers acquiesced.

In 1928, within weeks of moving to DuPont, Carothers decided to prove his theory about the bonding in giant molecules by building one. One of the best-known reactions in organic chemistry involves creating compounds called esters by joining together certain acids and alcohols. Carothers hypothesized that molecules that had acid functions on both ends could be reacted with molecules that had alcohol groupings on both ends in order to form long chains. He was right: Carothers had invented polyesters. Unfortunately, these early polyesters melted at a very low temperature and were impractical for fabrics. People were not interested in clothing that they couldn't wash in hot water or iron. But Carothers knew that organic acids also formed bonds with compounds called amines to generate amides, and that these bonds were particularly strong. In 1935, with the help of his longtime assistant Julian Hill, Carothers made the first heat-resistant polyamide that could be drawn into a fiber. It was almost like drawing a rabbit from a hat. DuPont even considered naming the new material "Dooparooh," for "DuPont pulls a rabbit out of a hat." Some favored "norun," but in the end the company settled on "nylon."

Nylon was durable and resilient. Carothers was not. Throughout his life, he had suffered from periods of depression, and he even carried around a capsule of cyanide in case he needed it. He became convinced that after creating nylon he would never again have another idea to match it. So, in 1937, in that Philadelphia hotel room, he dissolved his cyanide in lemon juice and drank it. The man who gave us nylon, one of the best examples of "better living through chemistry," would not live to see the results of his work.

A RUBBER MATCH

It's hard to fight an effective war without rubber. Fan belts, gaskets, gas masks, and tires are critical to the war effort. In 1930, a certain young American army officer was well aware of this, and he welcomed his assignment: to search for alternative sources of rubber. During World War I, the U.S. had lost access to the rubber plantations of South East Asia, and this had brought home to people the fact that reliance on foreign sources was a dangerous business. The officer's task was to investigate the possibility of using the latex of the guayule plant, which grew freely in Texas, as an alternative source of rubber. Discovering that this was indeed a viable plan, he recommended that the plant be protected and reserved for emergencies. But his superiors ignored his advice.

Then came Pearl Harbor. Within weeks of the attack, the Japanese had advanced into the Asian rubber-producing countries, and the U.S. lost about ninety percent of its supply. A hastily appointed presidential commission reported that the very success of the Allied cause was at stake. Luckily, American ingenuity came to the fore, and by 1942 U.S. chemical companies were producing over 200,000 tons of synthetic rubber — twice the amount the Germans were cranking out. German scientists had begun research on synthetic rubber in the 1930s, because Germany had also learned its lesson during World War I, when the Allies had cut off its rubber supplies. Pioneering chemist Hermann Staudinger had offered his country a head start: he had proposed that rubber was a polymer, a giant molecule made up of repeating units called monomers.

As early as 1826, Michael Faraday had distilled rubber and identified a small molecule called isoprene as one of its decomposition products. By 1879, chemists treating isoprene with hydrochloric acid had produced the first synthetic rubbery materials, but they were unable to explain how this actually happened until Staudinger introduced the concept of polymers. Now it became clear to them that the key to synthetic rubber lay in joining isoprene units into long chains. But their attempts to do this ended in failure. So the Germans started experimenting with molecules similar to isoprene and finally found that a mixture of styrene and butadiene would yield a suitable rubbery copolymer when treated with a sodium catalyst. This "Buna-S" rubber (the name derives from butadiene, sodium [Na], and styrene) served Germany's needs. They produced massive amounts; slave laborers in an Auschwitz factory made most of it.

Making Buna-S was not a simple business, as American scientists discovered. The polymerization worked best when the monomers were suspended in a solvent in the form of an emulsion, very much like fat droplets are suspended in water to form homogenized milk. Emulsifiers were needed to prevent the tiny droplets from coalescing, and soap was an ideal candidate. After all, soap works by emulsifying oil and water. The scientists selected Ivory soap, because they considered it to be the purest available. But there was a problem. While the soap was an excellent emulsifier, it somehow inactivated the sodium catalyst. Victor Mills, a chemist working for Proctor and Gamble, had an idea. Maybe the problem was the small amount of perfume that the makers of Ivory mixed in to mask the soapy smell. Mills made a special batch of scentless soap and found that it did the job perfectly. Normally, such discoveries would have been tightly guarded as industrial secrets, but the 1940s was no ordinary time. President Roosevelt had created the Office of the Rubber Director under William Jeffers, and he asked rubber manufacturers to pool their resources. Petroleum had been the classic source of styrene and butadiene, but now scientists were finding methods of making butadiene and styrene from the alcohol produced by fermenting of grain, potatoes, and molasses. By 1944, the U.S. was producing 700,000 tons of synthetic rubber, far outstripping Germany. Victor Mills's discovery undoubtedly helped win the war.

Charles Goodyear would have been astounded by these developments. Just about a hundred years earlier, he had produced the first practical samples of rubber. But Goodyear did not invent rubber. South American native peoples were already using this exudate of the *Hevea brasiliensis* tree when the European explorers first arrived on the continent. Columbus described how they played games with rubber balls and even coated fabrics with the latex to make primitive galoshes. Europeans found few uses for the substance. But Joseph Priestley, the discoverer of oxygen, determined that it could rub pencil marks off paper, and he coined the term "rubber." Charles Macintosh sandwiched a layer between sheets of fabric and created the first raincoat. But rubber got hard in winter and soft and tacky in summer. Goodyear dedicated his life to solving this problem. He tried mixing everything he could think of with the tree sap, including soup and cream cheese. Financing his work was a constant difficulty. Goodyear even sold his children's schoolbooks to help fund his research. Then came a happy accident. He had mixed the rubber with sulfur and spilled some of the mixture onto a hot stove. When the rubber was cool, it still had its elastic properties, but it was no longer sensitive to temperature. This vulcanized rubber eventually took the world by storm, but Goodyear, who had believed that God had assigned him the task of curing rubber, never benefited, and he died in debt.

The use of both synthetic and natural rubbers has increased dramatically in recent years. So have allergies to rubber. Research into this received a boost when Everett Koop, the former U.S. surgeon general, developed an allergy to the elastic in his underwear. We now know that certain proteins, present in small amounts in the latex, are responsible. This has renewed interest in extracting rubber from the guayule plant, which does not appear to have allergenic proteins. Perhaps researchers should have listened when that army officer recommended the use of guayule back in 1930. They would have listened twenty-two years later, when that army officer, Dwight D. Eisenhower, was sworn in as president.

OOBLECK AND BEYOND

I know from personal experience that it's not fun to be covered with oobleck. It wasn't supposed to happen. When I smashed my hand into that bowl filled with a slurry made of cornstarch and water, the stuff should have solidified immediately instead of splattering all over the place. And it would have, too, had I trusted what the physicists say and not hesitated before making contact. But I'm a chemist.

Dr. Seuss didn't have physics or chemistry in mind when he wrote the children's classic *Bartholomew and the Oobleck* back in 1949. It's the story of the cranky King of Didd, who was dismayed that nothing but rain, snow, fog, and sunshine ever came down from the sky. The king ordered his magicians to cast a spell, and soon a green goop — oobleck — began to fall, covering the kingdom. Was the king happy? Of course not. You know that such wishes have a way of backfiring. Everything in the land turned sticky, and the royal subjects flopped and floundered about in the green goop. Only when the king admitted that he should have left nature alone did the oobleck dry up and disappear. It has, however, reappeared in many a children's science experiment book. Oobleck is made by mixing a pound of cornstarch with about fourteen ounces of water to yield a truly yucky, but intriguing, mixture.

When you pick it up, oobleck just oozes out of your hand. But a quick squeeze converts it into a solid, which will maintain its form until you release the pressure. What's going on here? The cornstarch is composed of tiny granules, very much like grains of sand. When the granules are moistened, water flows into the spaces between them and acts as a lubricant. The oobleck flows. If you suddenly apply pressure, however, the water squeezes out from the spaces between the granules, dramatically increasing friction. The oobleck now solidifies — at least until you allow water to flow back between the granules by easing the pressure. Now you see why punching a bowlful of oobleck should result in an unexpected effect. The substance should harden and not splatter. Alas, it takes a fair bit of courage to punch the gooey mess. Easing up before entry will make you look like the Pillsbury doughboy.

Those little cornstarch granules that make oobleck possible can also perform chemical magic of a much more practical variety. Way back in 1811, Russian chemist K.S. Kirchof discovered that cooking starch granules with dilute sulfuric acid produces a sweet syrup. He didn't understand how it worked, but now we know that the acid causes the long chains of glucose molecules that constitute starch to break down into smaller fragments. The smallest piece formed is glucose itself, which, along with maltose (two glucose units), is perceived by our taste buds as sweet. Longer fragments, made of more glucose units joined together, have the effect of thickening the solution. These longer molecules have numerous atomic groupings (known as hydroxyl substituents), which can attract and bind water molecules. They also intertwine with each other. The overall result is that the motion of water molecules is impeded and the solution turns viscous.

Today, we rarely make corn syrup using the acid process. That's because various fungi, *Aspergillus niger*, for one, produce enzymes that degrade starch in a more controlled fashion. If we desire a sweet syrup, then we have to increase glucose and maltose yields. To obtain thickness, we need many chains made of at least six glucose molecules. By using enzymes appropriately, we can meet these requirements. Corn syrup has myriad uses. Add a bit of maple flavoring and you have artificial maple syrup. A dash of red food coloring and some titanium dioxide produce great fake blood — the makers of the latest Dracula movie went through gallons of it. Add an enzyme extracted from some streptomyces bacteria and you've got one of the most widely used corn products: high-fructose corn syrup.

Fructose, commonly found in fruits, is chemically very similar to glucose, but it's sweeter. As the bacterial enzyme converts glucose in the corn syrup to fructose, the sweetness increases. Most soft drinks today are sweetened with high-fructose corn syrup instead of cane sugar, because corn production is highly predictable, and the syrup is cheaper than sugar. Corn is an easy crop to grow. In North America, the acreage devoted to it is second only to that given over to wheat. Its various derivatives are present in over three thousand grocery items, and most of our livestock is reared on corn. Indeed, we consume about three pounds of corn each day in the form of meat, poultry, dairy products, and assorted other foods. Take a look at the label on a container of pudding or a jar of peanut butter. Chances are you'll find cornstarch listed. In some cases, it's modified cornstarch. This usually means that manufacturers have added certain reagents — phosphates, for example — to make some of the starch molecules cross-link. This joining together of the molecules to form a lattice affects the translucence, texture, stiffness, and moisture-holding capacity of the product. Unmodified cornstarch yields a loose pudding, whereas a spoon will stand up in a pudding made with modified starch. As if all these uses for corn weren't intoxicating enough, consider that we can ferment corn mash to produce whiskey. And we can also use it to make gasohol.

How about plastics and fabrics made from cornstarch? The people at Cargill Dow, a chemical company in Nebraska, have found a way to convert corn into biodegradable plastics. They ferment cornstarch to produce lactic acid, which they then convert to polylactides — long chains of the acid. Presto! They have created a plastic that can be molded and spun into fibers, and it comes from completely renewable feed stocks. A wedding dress made completely from polylactides is already on the market. In this instance, people who insist that old-fashioned weddings are corny could be perfectly right.

The Eyes Have It

My first trip to New York was in 1964. A couple of buddies and I decided we had to see the World's Fair. When we arrived at the fair site in Flushing Meadows, we looked around and considered what attraction we should visit first. The longest line was in front of the Vatican pavilion, of all things. We joined it, figuring that all these people must know something. They did. Michelangelo's magnificent *Pieta* was the pavilion's centerpiece. For the first time in five hundred years, the Church had agreed to display the work somewhere other than Saint Peter's Basilica, on the condition that its hosts take elaborate safety measures — such as placing the statue behind Plexiglas panels.

Being forced to view this glorious work of art through plastic did not sit well with the art critics. It gave one "a feeling of violation," the critic for the *New York Times* said; the plastic made the *Pieta* look "helpless and cold." Strange . . . that's not the way I saw it. I marveled at the plastic: it was virtually invisible, yet so strong that it could withstand bullets. It was a plastic that could have altered the course of history if, just a year before, President Kennedy hadn't refused to ride in a convertible shielded with it on that infamous visit to Dallas.

Plexiglas, or polymethyl methacrylate (PMMA), belongs to that class of substances we call polymers. These are long molecules made up of repeating units (monomers), much like a chain of many individual links. The links in this case are molecules of methyl methacrylate. In the 1920s, Dr. Otto Rohm, a German chemist, was the first to find a way of converting the liquid monomer to solid, clear sheets of polymer. And it probably never would have happened if he hadn't developed an interest in dog poop.

When Rohm was working for the Stuttgart Municipal Gasworks in 1904, an unpleasant odor wafting into his office from a neighboring tannery would often annoy him. He knew that tanners softened hides by immersing them in pits of fermented dog dung, and he began to wonder whether there might be an alternative to this distasteful method. The smell was very much like that produced by some of the by-products of his own industry, so Rohm decided to investigate a synthetic alternative to canine excrement. He soon came up with one. "Oropon," as he called the synthetic tanning agent, was an instant success. Rohm formed a partnership with Otto Haas, a businessman who had immigrated to America from Austria, and the two established the Rohm and Haas Company, which would become one of the most successful chemical-manufacturing firms in history.

With money rolling in, Dr. Rohm was able to devote time to his real interest: chemical research. He had earned his doctorate in 1901 by submitting a dissertation on the chemistry of acrylics, interesting substances made of raw materials isolated from petroleum or natural gas. Now, in the 1920s, Rohm picked up the threads of this research and discovered a way to join molecules of methyl methacrylate to make polymethyl methacrylate. The clear sheets of Plexiglas, as they named the novel substance, had obvious commercial appeal. They were transparent and strong, and they could be heat-molded into whatever shape one desired. Dr. Rohm was soon sporting the world's first pair of glasses that had acrylic lenses instead of glass ones. This gave the Luftwaffe an idea. Why not replace glass windows in aircraft with shatter-resistant Plexiglas? An excellent proposition — but there was a hitch. It was impossible to produce methyl methacrylate, the required starting material, on an economically viable scale. The answer to this dilemma would come from across the Atlantic, from the chemistry labs at McGill University.

At McGill, William Chalmers had found a way to make methyl methacrylate from acetone and hydrogen cyanide, both of which were readily available. Chalmers knew that John Crawford, a chemist at Imperial Chemical Industries (ICI) in England, was working with acrylic polymers, and he suggested to Crawford that he try the new method for making methyl methacrylate. Crawford successfully scaled up the process, making the mass production of polymethyl methacrylate possible. ICI called it Perspex. The Royal Air Force, like the Luftwaffe before them, recognized the potential of the material. They were in the process of developing the Spitfire fighter, and it featured a canopy made of plastic-reinforced glass that the pilot could draw over his head. Perspex, however, would clearly be better. By 1936, a Perspex manufacturing plant was in operation at Billingham, and Spitfires with Perspex canopies began to roll off the assembly line. Next came the B-19 Douglas Superbomber, in which the bombardier compartment and the machine-gun turrets were made of Perspex, allowing their occupants unobstructed views.

Perspex, as used by the British and the Americans, and Plexiglas, as the Germans called it, was a tremendous improvement over glass, but it was not indestructible. Direct hits could shatter it and send tiny slivers flying everywhere. On occasion, slivers would lodge in a pilot's eye. Under normal circumstances, a foreign substance in the eye causes terrible irritation, but British eye surgeon Dr. Harold Ridley noted that Spitfire pilots did not suffer this reaction. Somehow, their eyes tolerated this particular foreign material. Ridley now had a vision. Maybe this was the key to curing cataracts, those opaque deposits that form in the lens of the eye as we age. At the time, the only method of treating cataracts was surgically removing the lens and fitting the patient with "Coke bottle" glasses, which did the job that the natural lens had done. The widespread belief was that any kind of implanted lens was doomed to fail because the eye would reject it. But maybe it wouldn't reject polymethyl methacrylate, Ridley thought.

In 1949, Ridley carried out his first successful Perspex implant. The plastic performed well, but the surgical techniques were not refined enough. Lenses would often slip out of place, and the trauma of the surgery led to all kinds of complications. Most of these problems were eventually solved by the Dutch ophthalmologist Dr. Cornelius Binkhorst, who had studied under Ridley. It was from Binkhorst, in 1967, that Montrealer Dr. Marvin Kwitko learned the fine points of lens implantation, and Dr. Kwitko went on to pioneer the procedure in Canada. Although people initially greeted the innovation with skepticism, by 1975, Kwitko had demonstrated the viability of lens implantation and had begun offering a training course. Eventually, hundreds of ophthalmologists from across Canada and the U.S. would take the course and learn the fine points of cataract surgery and lens implantation.

The progress we have seen in cataract surgery has been truly phenomenal. It wasn't very long ago that cataract patients had to lie for weeks in their hospital beds, surrounded by sandbags to prevent them from moving. Today, doctors perform cataract surgery on an outpatient basis, and it usually takes them no more than fifteen minutes. They emulsify the natural lens with an ultrasonic probe so that they can remove it through an incision so tiny it doesn't even require sutures. Then they insert the new lens through the same opening. Silicone has joined polymethyl methacrylate as a material for lens manufacture, and researchers are working on developing lenses that can focus both near and far, perhaps eliminating the need for reading glasses after cataract surgery.

Chemistry, ophthalmology, and a good dose of serendipity have allowed many seniors to regain clear vision. I wonder if that *New York Times* critic who objected to viewing the *Pieta* through Plexiglas is now looking at the whole world through polymethyl methacrylate. If he is, I suspect that he doesn't feel violated by the plastic lens.

FROM TORPEDOES TO AIRBAGS

The Virginia country road was narrow, and the visibility was poor. The Chrysler LeBaron pulled out to pass another car. It never made it: the LeBaron slammed head-on into an oncoming vehicle, also a Chrysler, with a terrifying crunch. Other drivers stopped and rushed over to the tangled heap of metal, fearing the worst. To their astonishment, both drivers crawled from their wrecked cars unhurt. The year was 1990, and for the first time ever, two vehicles equipped with airbags had collided.

The original idea for airbags was born in the fertile mind of none other than Leonardo Da Vinci. "Baghe di vento," or "bags of air," he called his invention, which he obviously hadn't designed for cars. He'd designed it for flying men. Or, at least, for men who were attempting to fly. Da Vinci was fascinated by flight, and he dreamed of various flying machines, yet he was realistic enough to consider the risks. Brave men who strapped on wings, he thought, should also strap on bags of air to protect themselves should they fall from the sky like rocks.

But it was a different kind of rock that inspired the modern airbag in 1951. This rock was sitting in the middle of a road, and John Hetrick swerved to avoid it, ending up in a ditch. He was thankful that his daughter, who was sitting beside him, was unhurt, but he couldn't help thinking about what would have happened if she had been thrown against the dashboard more forcefully. On the way home, Hetrick began dreaming of sponges and cushions that could offer protection in the event of a crash. A memory from his days as a navy torpedo technician popped into his mind. Hetrick recalled being directed to work on a torpedo in a maintenance shop. Torpedoes are propelled by compressed air. Suddenly, by accident, the compressed air in Hetrick's torpedo was released. This was of no great consequence, but one detail of the occurrence stuck in Hetrick's mind: the tarpaulin covering the torpedo had flown into the air.

Here was a possible solution to his crash-protection problem. Could he come up with a device that would fill a pillow with air in the event of a collision? Hetrick worked on the idea and constructed a prototype. In 1952, he was granted the first patent for what would become the airbag's predecessor. The original idea of using compressed air turned out to be unworkable, because the air cylinder itself posed a risk. What if it was damaged in an accident and took off like a rocket? Furthermore, car manufacturers of that era were more interested in enticing customers with huge engines and tail fins than airbags. But as the slaughter on the highways continued unabated, carmakers began to realize that something had to be done. They installed seat belts, but most drivers didn't use them. Then they gave serious consideration to airbags, but how could they construct a bag that could inflate within a few milliseconds of impact without using compressed gases?

They found their answer in a fascinating chemical called sodium azide (NaN_3). When ignited by a spark, this substance releases nitrogen gas, which can instantly inflate an airbag. The only problem was that the reaction also forms sodium metal, which reacts with moisture to generate sodium hydroxide, a highly corrosive substance. Thus, a burst airbag could wreak havoc. At this point, chemical ingenuity came to the fore. If they included potassium nitrate and silicon dioxide with the sodium azide, then the only products that would form in addition to nitrogen would be potassium silicate and sodium silicate. Both of these are inert, harmless substances.

An airbag is designed to release some gas just after it deploys; this helps cushion the body against impact. Hitting a fully inflated, unyielding airbag would be catastrophic. Before the carmakers started promoting their new protective device, they had to ascertain the safety of its contents. In the 1970s, Mercedes settled this issue by putting a cage full of canaries in a car and deploying an airbag. Canaries are extremely sensitive to toxic gases, but the birds survived the experiment. By the late 1980s, airbags had become a common feature in automobiles. They have since saved thousands of lives.

But, as there is with any scientific advance, there is a "but." Airbags are not problem-free. While the chemistry involved in curbing sodium hydroxide production is clever, it isn't foolproof. Airbags have released small amounts of the caustic material; in rare cases, this has caused severe eye injuries, even blindness. The most serious concern, however, is the damage that can be done by an airbag rocketing out at an astounding speed of up to 330 kilometers per hour. Those who sustain a blow to the head under these circumstances risk death. Tragically, over one hundred people, mostly children and small adults, have been killed in this fashion; some lost their lives in low-speed collisions that would not otherwise have been lethal. Many researchers are looking for ways to ensure that airbags inflate only when necessary, and that they are deployed in the safest possible way.

They are exploring the use of sensors that can gauge a person's weight, allowing a computer to calculate how and whether an airbag should be inflated. In any case, small children must never be allowed to occupy the front seat of an airbag-equipped car. Some researchers even argue that seat belts afford better protection than airbags.

There is a further problem that we need address. Sodium azide is more toxic than cyanide. What will happen to all the azide in cars headed for the junkyard? What if their azide canisters are not removed? If sodium azide is released, it can react with water to form hydrazoic acid, which is not only toxic but highly explosive as well. Sodium azide also reacts with metals such as copper or lead to form explosive copper or lead azides. Just ask the plumbers who were called to a lab where workers had been using sodium azide solutions. When they removed a piece of copper pipe and tossed it into the garbage, it exploded. An unfortunate and shocking way to learn about the chemistry of azides!

Soap Story

The Burlington Arcade in the heart of London is a shopper's dream. In this fascinating place, you'll see one-of-a-kind antiques, piles of cashmere sweaters, and unique jewels glistening in showcases. Mere mortals can't actually afford to buy much there, but I was determined to come away with something. My chance came in a soap shop. That's right, a soap shop — a shop that sells nothing but soap. And this is no ordinary soap; it's special, handmade, designer soap. I thought that some of this wonderful soap would make a good souvenir, since soap making is one of the oldest chemical processes known to humankind.

According to Pliny, the Roman historian, the Phoenicians fabricated the first soap in about 600 B.C. from boiled goat fat and caustic wood ashes. The ashes served as a source of sodium and potassium hydroxides (lye), which reacted with the fat to produce soap. A more captivating story links the discovery of soap with Sapo Hill on the outskirts of Rome, a site where ancient Romans sacrificed animals to heathen gods. To do their washing, local women used clay they found at the foot of the hill, on the banks of the Tiber, because of its remarkable cleaning properties. Fat from the sacrificial animals had apparently reacted with hot wood ash to form soap; the substance was washed down to the riverbank and absorbed by the clay. An enterprising American company now manufactures soap "the old-fashioned way" under the name Sappo Hill. The only sacrifice involved here is made by the consumers who open their wallets wide to buy the stuff.

Fats are composed of a backbone of glycerol linked to three fatty acids, and we refer to them, accordingly, as triglycerides. Lye breaks the linkages, liberating glycerol and forming species known as the sodium or potassium salts of the fatty acids. These are soaps. They are characterized by a long tail of twelve to eighteen carbon atoms and a head that features two oxygens that have grabbed the sodium or potassium ions furnished by the lye.

Soap has a double-barreled cleaning action. The tail is soluble in oil, and the head is soluble in water. Most dirt is of an oily or greasy nature, and it attracts the tail, leaving the head to be anchored in water. Rinsing then pulls the oily dirt off whatever surface it is attached to. Soap also changes the wetting ability of water. It may sound bizarre to talk about the wetness of water, but, in a scientific sense, not all water is equally wet. Wetting is a measure of the ability of water to spread. Normally, water on a surface will form beads. This is because water

molecules are attracted to each other more strongly than to a surface or to air. But if we dissolve soap in the water, then its molecules will gather at the surface, since water repels the long hydrophobic (water-hating) tail. This reduces the attraction between water molecules at the surface and allows for easier spreading. The water can now wet a surface more thoroughly and clean it more efficiently.

The Romans and the Phoenicians may have had soap, but its use declined with the rise of the Christian Church, which warned its followers of the evils of exposing the flesh, even to bathe. Only as Europe emerged from the Dark Ages did people's thoughts turn to cleanliness. Still, soap was only for the filthy rich. A friend of Queen Elizabeth boasted that she "hath a bath every three months whether she needed it or no." Soap was very expensive; this was due partly to the cost of manufacture but mostly to the taxes imposed upon it in 1712, during the reign of Queen Anne. These taxes remained in effect for 150 years, and it was a serious business. No one could manufacture soap without having a taxman on the premises; and every batch had to be completely accounted for. Some candle makers tried to cheat the taxman by making soap on the side, but the government got wind of this and deemed it illegal for candle makers to possess lye.

Today, we can make soap very cheaply, thanks to Frenchman Nicolas Leblanc, who, in 1879, discovered a method of making lye from brine. Yet Leblanc's breakthrough certainly isn't reflected in the prices they charge for soap at Immaculate House in the Burlington Arcade. You can invest your annual salary in their lavender, marmalade, or flower garden soap. The little scroll included with the soaps reveals that they are all made from palm, coconut, olive, and sunflower oils, which have been reacted with sodium hydroxide, or lye. Just like any other soap. But, of course, there are the extras: the flower garden has marigolds and roses; the lavender is studded with lavender flowers; and the marmalade — well, that contains little bits of orange peel. Sounds great for washing out a dirty mouth. And how well do these soaps work? As well as any other. Truth be told, there is not much difference between soaps. That's why marketers have to resort to gimmicks. Some soaps have added fat or lanolin, and they tout their moisturizing effects. Others leave in lots of glycerol, which makes them transparent. "Gentle to the skin!" scream some labels. Indeed, some highly alkaline soaps may strip the skin's natural oils and cause more irritation than products that have a pH (measure of alkalinity or acidity) close to that of water. By and large, though, soap is soap. Find one you like and use it often. Especially on your hands. Frequent washing undoubtedly reduces the risk of microbial illness. No special ingredients needed.

I hope I've managed to clean up some of the confusion surrounding soap. It's a necessary task. I know, because I once asked some elementary school students if anyone knew what soap was made of. One bright youngster confidently blurted out "Ivory!" But, judging by the soap prices at the Burlington Arcade, I cannot definitively rule this out.

DETERGENTS AND DROWNING FLEAS

The performing fleas were among the highlights of P.T. Barnum's traveling circus. The tiny creatures demonstrated their strength by drawing carriages 131,000 times their own weight; they also played music and walked on water. Now, walking on water is not so hard if you're a flea. And it has nothing to do with fleas being less dense than water. They're not. It has to do with one particular property of water: surface tension.

Water is made up of molecules in which two hydrogen atoms are attached to an oxygen atom. Everyone knows that. But not everyone knows that there is also an attraction between the hydrogen atoms of one molecule and the oxygen of another. These hydrogen bonds, as they are called, are quite weak, yet they have a significant effect on water. Attraction between adjacent molecules causes the water surface to develop a virtual skin. Just try to push your finger slowly into a glass of water. The surface will actually bend before the finger punctures it. Or try placing a paper clip — clearly heavier than water — on the water surface. If you do this carefully enough, the paper clip will float. And so will a flea.

Surface tension may permit fleas to walk on water, but it creates a real problem for us when it comes to washing. Ideally, for cleaning purposes, we would like water to flow unimpaired into the nooks and crannies of whatever material we are trying to wash. But water has a resistance to flow, a phenomenon that is apparent when we observe a drop on a glass surface. Instead of spreading freely, the water forms a bead. To allow the water to spread, and thus enhance cleaning, we must decrease the attraction between adjacent water molecules. This is where surfactants come in. These consist of molecules that interfere with the attraction between water molecules and actually increase the ability of water to wet a surface.

The earliest example of a surfactant is soap. Soap, as we have seen (page 200), is not only great at reducing surface tension, but it also binds dirt to water. It consists of long molecules, one end of which is soluble in water, the other in oil. So, one end of the molecule anchors itself in the oily dirt, and the other end binds to water, enabling us to rinse the dirt away. The difficulty with soap, though, is that it does not work well in hard water — that is, water that has calcium or magnesium ions dissolved in it. Soap reacts with these ions to form a precipitate, or scum. This problem triggered the search for synthetic surfactants that would have the properties of soap but would not precipitate out in the presence of minerals.

In the 1930s, chemists rose to the challenge and developed the first synthetic detergents, known as branched alkyl benzene sulfonates; they were made from petroleum products. This solved the precipitation problem, but they still had to address another difficulty. Sulfonates did not come out of solution when calcium and magnesium were around, but they still reacted with them. Although the sulfonates stayed in solution, their cleaning ability was compromised. Enter the detergent "builders," first introduced in 1947 by Proctor and Gamble, in Tide.

Chemists looked for a reagent to add to detergent that would somehow bind the minerals in water, leaving the detergent free to do its job. Phosphates seemed to fit the bill. They were non-toxic and cheap. Manufacturers loaded up their detergent formulations with phosphates; but then, in the 1960s, a new problem cropped up. Rivers and lakes began to develop a strange foam residue, and sometimes suds even emerged from kitchen taps. Alkyl benzene sulfonates, it turned out, were not bio-degradable, and they survived their journey through sewage treatment plants. So the chemists went back to the lab to tinker with the structure of the branched alkyl benzene sulfonates and came up with "linear alkyl benzene sulfonates," which were susceptible to attack by microorganisms. They had solved the foaming river problem.

But the chemists couldn't relax for long. Blooms of algae replaced the foam in natural waters. Phosphorus is an essential nutrient for plants; it's one of the three elements, along with nitrogen and potassium, that we incorporate into fertilizer. The algae blooms indicated that water plants were being fertilized. At first, this didn't seem to pose any difficulty — after all, water plants photosynthesize and produce oxygen, which aquatic life needs. But when the fertilized water plants died, their degradation used up oxygen, and the net result was a reduction of dissolved oxygen and an impairment of aquatic life. One solution was to design sewage treatment plants where phosphates could be removed; the other was for detergent manufacturers to cut back on phosphates and replace them with other builders.

Minerals known as zeolites (hydrated alkali aluminum sili-cates, for the chemically inclined) are an alternative to phos-phates. They are less efficient at binding calcium and magnesium, so we have to use more of them. And their production process is not exactly pollution-free. Recently, researchers have raised another concern about detergents. This one hits below the belt. They have discovered that breakdown products of phenol ether sulfates (a popular detergent) have estrogenic properties. In fact, they are related to nonoxynol-9, a common ingredient in spermicides. Some scientists even link the reduction in average global sperm counts with certain detergents.

We must address this situation, but I don't think anyone is suggesting that we give up our detergents. They are just too useful. Why, you can even use them to do away with fleas. The little guys are performers at heart. Place a dish of water with a light shining on it in the room where you suspect they are hiding, and add a touch of detergent to the water. The fleas will go for the spotlight and jump towards the "stage." But because the detergent has reduced the water's surface tension, there will be no walking on water. Since swimming is not one of the talents that fleas possess, they will sink and drown.

UNTANGLING THE WEB OF SPIDER LORE

I really don't understand what Little Miss Muffet's problem was with spiders. There she was, on her tuffet, happily eating her curds and whey, getting a good dose of calcium, when along came a spider, sat down beside her, and scared her away. It wasn't even a black widow or a brown recluse, in which case a little trepidation may have been in order. It was a common house spider. How do we know this? Because that's the species her father kept around the house.

Oh, yes, Little Miss Muffet was a real person. Her real name was Patience, and she was the daughter of Thomas Muffet, a sixteenth-century British physician who kept spiders because he liked them to decorate his rooms with their tapestry. His patients must have thought he was a little bizarre, and certainly Patience had no patience for the arachnids. That's probably because she knew so little about them.

A bite from a black widow or a brown recluse spider can be life-threatening, but the truth is that you stand a greater chance of being attacked by a shark or struck by lightning than sustaining such a bite. Spiders are actually beneficial. They kill more insects than all birds combined, and far more than insecticides. Each year, spiders eat a quantity of insects that exceeds the weight of the total human population. Now there's a statistic to chew on!

In the 1970s, Chinese farmers began to harness spider power. They discovered that they could employ the eight-legged predators (don't call them insects, because that's not what they are) to patrol cotton fields and protect the plants from insect infestation. They house their wolf spiders, jumping spiders, and crab spiders (which paralyze insects with poison instead of ensnaring them in webs) in "spider motels" (straw bundles) in the fields. The arachnids hibernate happily. They wake up in the spring feeling ravenous and ready to gorge on insects. In some areas, pesticide use has decreased by sixty percent. The Australian funnel-web spider may also reduce our reliance on insecticides. One of the compounds in its venom is deadly to cockroaches, crickets, and fruit flies, but it's harmless to mammals. We may one day be able to produce it through genetic engineering techniques in quantities large enough to combat insects.

Banana spiders live in tropical climates and release a scent that lures cockroaches to their doom. It's safer than roach sprays. Or, if you have moths, how about some bola spiders? They dangle a strand of silk with a drop of glue at the end and swing it around to catch their prey. Bolas emit an odor that resembles a female moth's sex attractant, and it is very enticing to male moths. They come looking for love and end up as spider food.

Perhaps even more interesting is the ogre-faced spider that throws a web over its prey — mainly ants — and lifts them off the ground so they cannot leave a scent as a warning to others. Then, once the web has done its job, the spider eats it and recycles the protein. How's that for environmental friendliness?

Then there is the clever Central American spider that disguises itself as an ant by holding a pair of its legs over its head to mimic antennae. So disguised, it climbs into an ant nest and has a feast.

And how about the male European crab spider? Now there's a kinky little fellow! During courtship, he spins a veil-like web and uses it to tie up the female. Good thing he does, because the female has a nasty habit of eating the male after mating. But, if the male spins his web properly, he can come and go by the time the female wriggles free.

Not surprisingly, drugs affect a spider's skill at spinning its web. That's why NASA scientists think spiders could replace other animals in chemical-toxicity testing. On amphetamine, the spider spins its web fast, but without much planning, leaving large holes. On chloral hydrate, its efforts are spindly and ineffective. If a spider is high on marijuana, the web starts out normal, then loses its pattern; the spider just gets too relaxed and gives up. On caffeine, it strings a few threads together at random. The more toxic the chemical, the more deformed the web. Perhaps, with the aid of a computer program, researchers can quantify the effects and produce an accurate test for toxicity.

You never know when spiders are going to come in handy. A seventy-year-old lady thought she felt a spider bite her, and then she noted red dots on her skin. She started to feel sick, and she consulted several doctors. None of them could diagnose spider bite, but one did detect breast cancer. That spider may have saved the lady's life. Spiders may save other lives as well. During a stroke, the brain releases a substance called glutamate, and this is responsible for some of the ensuing damage. Researchers have isolated glutamate antagonists from spider venom, and these may be useful in protecting brain cells from damage after a stroke.

Maybe if I'd had the chance to sit down with Patience Muffet and talk to her about spider science, I could have prevented her panic. For in this instance — and in so many other areas of life — ignorance breeds fear. Next time, instead of attacking spiders with chlorpyrifos, cypermethrin, bifenthrin, neem oil, or even a broom, sit down — on a tuffet, if one is handy — and think about how amazing these little creatures are. Then give a thought to how they suspend their webs across what appear to be unbridgeable distances. Maybe you can find more information about this on the World Wide Web.

A Bolt out of the Blue

Maybe it all started with lightning. Life, that is. At least that's what a landmark experiment carried out by Harold Urey and Stanley Miller in 1953 implied. The two scientists placed a mixture of water vapor, hydrogen, methane, and ammonia in a flask equipped with a pair of electrodes capable of generating a stream of sparks to simulate lightning. Most people believed that these gases were the major components of the primordial atmosphere, and Urey and Miller wanted to see whether an electric discharge would cause them to react and form the molecules that characterize life.

After a couple of weeks of flashes, they were rewarded. The contents of the flask turned cloudy and brown. Something had obviously happened. Chemical analysis showed the presence of amino acids, molecules that are the components of one of the most important types of biomolecules: the proteins. Follow-up studies revealed that proteins themselves, along with a number of other organic compounds, can be generated in such a system. Urey and Miller's experiments fell far short of explaining how such molecules could have organized themselves into living cells, but they did suggest a possible role for lightning in creating life.

During a summer thunderstorm, however, we are more concerned about lightning taking life. Indeed, Benjamin Franklin was a very lucky man when, in 1752, he flew a silk kite during a storm, hoping to demonstrate that lightning was a form of electricity. He had tied a key to the kite string, and he noted that "The electric fire would stream plentifully from the key on approach of your knuckle."

The kite had developed a static charge. It's fortunate that lightning never struck it; if that had happened, then Franklin's epitaph, "He tore the lightning from the heavens, and the scepter from the hands of tyrants," would not have included the second phrase. The great man would never have had the chance to play a pivotal role in the War of Independence. He would have been fried, like those foolhardy individuals who have since attempted to duplicate his classic experiment. Heat from lightning can exceed a temperature that is far hotter than the surface of the sun.

Franklin had actually begun experimenting with electricity six years earlier, after witnessing a demonstration of static electricity performed by a Dr. Spencer of Edinburgh. He immediately purchased Spencer's equipment — a primitive version of the apparatus that makes kids' hair stand on end, the one that we see today in virtually every science museum. At the time, static electricity was still a mysterious phenomenon, even though people had been observing it for thousands of years. The ancient Greeks observed that if they rubbed a piece of amber against cat fur and then tried to touch another object with it, then the amber and the object would repel each other. Indeed, our word *electricity* derives from "electro," Greek for amber.

Before Franklin, demonstrations of static electricity were like parlor tricks. For example, in 1730, Stephen Gray achieved a degree of fame by suspending a boy on two silk ropes, rubbing his feet with a ball of sulfur, and drawing sparks from his face. Franklin studied such phenomena, along with lightning, concluding that we could explain them all as the result of the transfer of an "electric fluid" from one substance to another. Rubbing two objects together left one object with an excess of the fluid and the other with a deficiency of it. Pretty inventive stuff. Not too far from the modern explanation that charges are created by the transfer of negatively charged electrons from one material to another. The substance that loses electrons becomes positively charged; the one that gains them takes on a negative charge.

When we touch a substance that has an excess of electrons with a conductor, we may cause the electrical charge to be drained away. If we just bring the conductor near but do not touch the charged object with it, then electrons may jump the gap and generate a spark. This can be a minor problem, like the one we encounter when we have to touch a doorknob after walking across a nylon carpet on a dry winter day; or it can be major problem, like the one experienced by some IRA terrorists of the past.

Apparently, static electricity contributed to the deaths of several terrorists. Their job was to transport explosive devices by car; these devices were made from ammonium nitrate fertilizer and diesel oil, or from weed killer (sodium chlorate) and sugar. Their clothing, made of synthetic fabrics, would rub against the polyisocarbonate covering of the car seats, building up a large static charge. (This happens easily, because car tires provide good insulation from the ground.) Then these charged terrorists would come close to a conductor, such as a metal bomb casing, and . . . BOOM!

Had they attached a wire to the car that dragged on the ground, providing a path for the electrons to escape, they would have prevented the buildup of static electricity and spared themselves. Benjamin Franklin recognized this phenomenon after his kite experiment and suggested that people use lightning rods to protect their buildings. Some inventive designers even created hats featuring lightning rods connected to a wire that trailed behind the wearer.

It is hard to believe, but many people at first resisted using lightning rods. The Church opposed the idea: lightning was divine intervention, a response to some mortal sin, and we should not interfere with it. In early America, when lightning struck a house, firefighters would not attempt to put out the flames; they would only douse neighboring houses to prevent the fire from spreading. Gunpowder was commonly stored in churches because churches supposedly had divine protection. But, in fact, they didn't; what they had were pointed steeples that drew lightning. Finally, in 1767, when a Venetian church exploded, killing some three thousand people, religious opposition to lightning rods disappeared.

Franklin himself had felt the wrath of the pious multitudes. In 1756, he was attacked by ministers who thought he was responsible for an earthquake in Boston. The ministers charged that Franklin's lightning rods had stolen lightning from the Almighty and channeled it into the ground, where it had created the earthquake. Franklin cleverly retorted that protecting oneself from lightning with a lightning rod was no more sacrilegious than protecting oneself from rain with a roof.

Since today everyone regards such theological concerns as outdated, I can offer advice about protecting yourself against lightning without being accused of blasphemy. Here it is. Do not stand under a tree, because trees are about twenty percent moisture, and humans about sixty-five percent. Water conducts electricity very well, and the striking bolt will follow the path of least resistance; if it hits a tree, the bolt will jump from the tree to anyone standing under it before passing to the ground. Tents, golf carts, hills, swimming pools, lakes, and open spaces are unsafe. A car with closed windows offers excellent protection, because metal will conduct electricity to the ground. Contrary to popular belief, the protection has nothing to do with the rubber tires. A bolt of a few million volts is not going to be stopped by an inch of rubber.

If you're stuck outdoors, your best bet is to crouch down with your heels together; do not lie on the ground. If you lie down and a bolt strikes nearby, the electricity may pass through your whole body. Not a happy event. Indoors, get out of the tub or shower, stay away from appliances, hang up the phone, and turn off the TV. Even if you're bored, do not go outside and fly a kite.

LOOKING BACK

A Fair to Remember

I had a strange lunch the other day. A hot dog, a cracker with peanut butter, a wad of cotton candy, and an ice cream cone, all washed down with Dr. Pepper. No, I haven't taken leave of my senses. I indulged in this nutritional fiasco out of a sense of history. These items, you see, have something interesting in common. They were all introduced to the general public at the same time and at the same place. The time was 1904, and the place was the St. Louis World's Fair.

St. Louis can be brutally hot in the summer. For its inhabitants, ice cream has long been a popular cooling treat. In 1904, the stuff wasn't new — Americans had been savoring it for over a hundred years. One day, one of the city's fifty or so ice cream vendors ran out of the little dishes in which they served the frozen delight in. A neighboring Syrian pastry concessionaire came to his aid. Ernest Hamwi suggested that a rolled-up waffle would hold a ball of ice cream quite nicely. Indeed it did. The waffle also eliminated the need to wash dishes. The inventive Hamwi liked the idea so much that he founded the Cornucopia Waffle Company, which began producing waffle cones for ice cream by the millions.

While Hamwi popularized the cone, he didn't actually invent it. In 1903, Italo Marchiony, an Italian immigrant who sold ice cream from a pushcart in New York City, had filed a patent for "a pastry cone to hold ice cream," but he'd failed to capitalize on his idea. So what's a cone made of? No big secret: flour, water, cornstarch, sugar, a little fat, soy lecithin (to prevent the fat and water from separating), ammonium carbonate (to release carbon dioxide during baking, yielding an airy texture), sodium metabisulfite (a preservative), salt, and coloring.

Ice cream was not the only sweet concoction available to World's Fair visitors. There were plenty of cotton-candy stands about. A couple of Tennessee candy makers had invented an electric machine that was essentially a spinning, heated bucket with many small perforations in it. They poured sugar into the bucket; the sugar melted, and the rapid rotation of the bucket forced it through the tiny holes. As soon as the melted sugar hit the cooler air outside the bucket, it froze into streamers. The candy makers sold their Fairy Floss in little wooden boxes for the hefty price of twenty-five cents. Those customers who waited too long after purchase to get their sugar fix were disappointed. Sugar picks up moisture from the air very quickly, and the Fairy Floss would collapse as the sugar dissolved. That's why today most cotton candy is sold in plastic bags. Heating the sugar can also be problematic, because the candy maker must take care to avoid caramelization — brown cotton candy does not have much appeal. Brown peanut butter, however, does. And that was also a hit at the St. Louis World's Fair.

Peanut butter had existed since 1890, but few people knew about it. It was developed by a St. Louis physician who worried that some of his patients were not getting enough protein. In fact, they had such bad teeth that they couldn't chew meat. So the good doctor experimented with grinding peanuts into a buttery paste, which his patients happily gummed down. Dr. John

Harvey Kellogg, of cereal fame, took out the first patent on peanut butter in 1895. And one of Kellogg's former employees developed an efficient hand-operated peanut grinder. Now the world was ready to experience this gastronomic breakthrough. In 1904, it did just that. By then, C.H. Sumner had realized that one could greatly improve peanut butter's taste by roasting the peanuts before grinding them. We still do it that way today. But we also take measures to ensure that the oil doesn't separate from the solids in our peanut butter. Producers often add hydrogenated fats to increase the density of the oil, thus preventing it from rising to the top. Hydrogenated fats have a deservedly bad reputation, because they contain trans-fatty acids, which drive up blood cholesterol levels. But not as much as the fat in hot dogs, and there were plenty of those consumed at the World's Fair.

Unlike peanut butter and cotton candy, wieners and frankfurters were very popular before 1904. Customers knew what they were getting from Anton Ludwig Feuchtwanger, a Bavarian peddler who had managed to set up a little food stand at the fair. He handed each customer a cotton glove with which to hold the greasy frank. But so many of them went off with Feuchtwanger's gloves that he had to look for another solution to the greasy-hand problem. His brother-in-law, conveniently, was a baker. It was he who had the idea of sliding a frank into a bun, and an American tradition was born. What better beverage was there to wash the fatty dogs down than the popular new drink that was the talk of the town?

Charles Alderton, a young British pharmacist, invented Dr. Pepper in 1885. He was working at Morrison's Old Corner Drug Store in Waco, Texas, where he doubled as a soda jerk. Here he put his chemical talents to use and came up with a blend of fruit juices (contrary to the rumor, prune juice is not one of them), which he flavored with sugar and colored with caramel. His customers liked the taste, and one of them suggested that he name it after Dr. Charles Pepper, the Virginia doctor who had given drugstore-owner Morrison his first job. At least that's one version of the story. Morrison eventually marketed the beverage, which hit the big time after massive exposure at the World's Fair.

Now you see that there was at least some method behind my madness as I dined on these seemingly unrelated delicacies. And do you know what? After that bizarre lunch, I didn't feel like eating anything all night. Maybe there's a market out there for the St. Louis World's Fair Diet. Just kidding.

Lydia Pinkham to the Rescue

A small, modest gravestone in Pine Grove Cemetery in Lynn, Massachusetts, bears the simple inscription "Lydia Pinkham: 1819-1883." A visitor would hardly guess that it marks the final resting place of the most famous woman in North America during the latter half of the nineteenth century.

Lydia Pinkham wasn't born into fame. She achieved it by finding a niche that needed filling. She became an expert on "female problems" at a time when many women were reluctant to consult male physicians about such things. Lydia was a nondescript schoolteacher who dabbled in phrenology, the bizarre study of the relationship between bumps on a person's head and personality, and she prepared homemade remedies for various ills. One of her specialties was a concoction for female complaints — Lydia Pinkham's Vegetable Compound — which she brewed for friends and neighbors. We probably would never have heard of this potion if Lydia's husband hadn't lost the family's money through real estate speculation in 1873. The couple had to find some way to make a living. Why not try selling Lydia Pinkham's Vegetable Compound to stores? Son Dan went out to drum up business, while the rest of the family stayed home and brewed a big batch. Sales were sluggish until Dan had an idea. What the compound needed was an instantly recognizable bottle design. In that era, the makers of patent medicines competed for customer attention by using elaborate product labels and making outlandish claims. Dan opted for an image of simplicity and friendliness: his mother's face.

Soon, Lydia Pinkham's smiling grandmotherly visage graced not only the bottle but also the ads that the family took out in newspapers across the country. "Trust me," the sympathetic face seemed to say. And people did. Not only did they trust, but they also bought. I doubt that Lydia Pinkham's Vegetable Compound did much for "nervous prostration" or "prolapsed uterus," but it certainly solved one female problem: it cured Lydia Pinkham's financial ailments. Lydia Pinkham's Vegetable Compound made her the first millionairess in America.

What were people actually buying? A collage of licorice, chamomile, pleurisy root, Jamaica dogwood, life plant, dandelion root, and black cohosh. This last ingredient, derived from the rhizome of the *Cimicifuga racemosa* plant, already had a long history of use for "women's problems." American Indians valued it highly, drinking a brew they made by boiling the root in water. Its name derives from the Algonquin word for "rough," because of the texture of its gnarled root.

Black cohosh, like virtually any other plant, contains compounds that at a high dose have physiological effects. Because black cohosh has a history as a treatment for menstrual problems, researchers have investigated its estrogenic potential. Their experiments have shown that an alcohol extract of the plant does contain compounds that bind to estrogen receptors in the uteruses of rats, but we lack human data. However, plenty of anecdotal evidence exists — women claim that the substance has rid them of excessive bleeding, irregular periods, delayed menstruation, severe menstrual cramps, or menopausal hot flashes. Menstrual cramps are apparently relieved by the antispasmodic effect of black cohosh; it reduces the intensity of uterine contractions.

The ingredient in black cohosh that has estrogenic effects is formononetin. Researchers rationalize that if a woman lacks estrogen, as many women do in menopause, then formononetin will stimulate her estrogen receptors. But when a woman has too much estrogen, causing cramping, formononetin can block the activity of her body's natural estrogen. Along with other compounds in black cohosh, formononetin has been shown to reduce the levels of luteinizing hormone, which is linked to hot flashes.

At least one controlled clinical study suggests that black cohosh works. Researchers studied a group of women, all under forty, who had undergone hysterectomies. They gave some subjects estrogen supplements and others black cohosh to determine the effects of these substances on the women's premature menopause. Black cohosh was as effective as the estrogen supplements, but it took longer to kick in.

Researchers have also investigated other potential uses for black cohosh. Its antispasmodic effect may be useful in treating sprained muscles as well as arthritic pain, which is made worse by muscle tightness. Black cohosh also has some sedative properties, and it may work against coughs.

There are some caveats, though. Nobody knows what effect black cohosh will have on pregnant women or those suffering from breast cancer. Nobody knows if it interferes with estrogen replacement therapy or with the birth control pill. Nobody has studied its long-term effects. It may intensify the effect of blood-pressure-lowering medications. And there is the question of dose. The current opinion is that for menstrual cramps, three to four capsules containing forty milligrams of root, standardized to 2.5 percent triterpenes, is appropriate. The triterpenes are not active ingredients, but we can measure them with relative ease. Since their concentration is proportional to that of the estrogenic components, they provide a convenient means of standardization. The limited number of studies that have been done indicate that the dose I just described decreases the frequency of hot flashes. Some generally available supplements — such as Remifemin — contain black cohosh. Remifemin is probably the best-studied preparation, and it comes in liquid form (the dosage is forty drops, two times a day) or tablet form (two tablets, twice a day). While some women experience an alleviation of menopausal symptoms in as little as four weeks, most have to wait six to eight weeks for the same result. If Remifemin hasn't helped in twelve weeks, it is unlikely to do so.

Yet another ingredient of Lydia Pinkham's Vegetable Compound may have helped ladies feel better: a hefty dose of alcohol. The original version had close to twenty percent alcohol by volume, and it delivered an "unusual degree of satisfaction." And Lydia delivered satisfaction in another way as well. Ads and product labels invited women seeking advice on female matters to write to Lydia Pinkham. "Any woman," the ads maintained, "is responsible for her own suffering if she will not take the trouble to write to Mrs. Pinkham for advice." They wrote and wrote. And she wrote back. Lydia answered the letters herself, usually recommending an increased dose of her tonic. She was a prolific letter writer. Even death did not deter her. Although Lydia died in 1883, for years afterwards the letters that her loyal customers wrote to her continued to be answered — by "Mrs. Pinkham" herself! Maybe that vegetable compound really did have miraculous effects. There is one miracle it really did produce: longevity. Lydia Pinkham's

Vegetable Compound is still around. But don't look for black cohosh in the modern version. Current hot sellers like vitamin E, vitamin C, and iron have replaced it. Don't look for the twenty percent alcohol either. That's been cut way down. There's not much evidence that the present incarnation provides any benefit, but we cannot doubt that in its time, Lydia Pinkham's Vegetable Compound made many women happy. And, according to a little song popular in the early 1900s, Lydia may have done more than communicate with her correspondents from beyond the grave:

Poor Lydia died and went to heaven.
All the church bells they did ring.
But she took along her vegetable compound,
Hark, how the Herald Angels sing!

I bet they did. And it wasn't because of the black cohosh.

MAUVING ON

The National Portrait Gallery in London is an amazing place. It is so overwhelming that many visitors don't know where to turn first. Should they search for the only known authentic portrait of Shakespeare? Or should they pass the time with Dickens, Henry VIII, or Isaac Newton? It's a tough call. But I knew where I wanted to go. I wanted to find the room that featured the nineteenth-century scientists. I had to see the original of the picture that I had shown so often in lectures. I had to pay homage to William Henry Perkin, the man who, in 1856, single-handedly changed the course of chemistry.

During the early years of Queen Victoria's reign, the Germans dominated the pursuit of chemical knowledge. Bunsen, Wohler,

and, particularly, Justus von Liebig were the leading lights. Britain had no comparable personalities, which Liebig underscored on one of his English lecture tours. "England is not the land of science," he insisted. "There is only widespread dilettantism; their chemists are ashamed to be known by that name because it has been assumed by the apothecaries, who are despised." (Indeed, in England pharmacists are still known as chemists, although I suspect they are no longer despised.) Liebig's remarks struck a cord with Prince Albert, the queen's husband, who was a strong supporter of scientific inquiry. England needed to get on the chemical bandwagon, he thought, and it needed an institution that would allow it to do so. So he sponsored the establishment of the Royal College of Chemistry, and the college invited August Wilhelm Hofmann, a former student of Liebig's, to be its director. And it was there, in 1853, that a historic meeting would take place between the German chemistry professor and fifteen-year-old William Henry Perkin.

As a youngster, Perkin had been keenly interested in art and the fledgling field of photography. But these pursuits would later take a back seat to chemistry. When he was about twelve, Perkin recalled, "a friend showed me some chemical experiments and the wonderful power of substances to crystallize.... My choice was fixed, and I determined, if possible, to become a chemist and I immediately commenced to accumulate bottles of chemicals and make experiments." Perkin asked his father to enroll him in the City of London School, which was the only school offering practical chemistry lessons at that time. These courses were not part of the regular curriculum, and interested students had to pay extra to be taught by Thomas Hall, the chemistry master who had studied under Hofmann.

Hall immediately recognized young Perkin's potential and arranged for him to enter the Royal College of Chemistry. Perkin Senior had his heart set on his son becoming an architect, and he did not approve of his infatuation with a subject that had no career potential. But he finally did agree to give his son the tuition money, an investment that would eventually pay very handsome dividends. Under Hofmann's tutelage, young Perkin began to experiment with coal tar, the mucky residue that coal left behind when it was heated in closed vessels in the absence of oxygen to produce gas for gaslights. Hofmann's interests had always tended towards the practical side of chemistry — his course syllabus included dyeing and bleaching, and extracting drugs from natural sources. But Perkin's interest was aroused by a lecture on the potential use of coal tar derivatives to make drugs that were then only available from natural sources. Quinine, extracted from the South American cinchona tree, was the only treatment for malaria. And there wasn't enough of it.

Hofmann had carried out a chemical analysis of quinine, noting that its composition suggested it could be made by somehow combining two molecules of allyltoluidine, a coal tar component, with oxygen. This idea intrigued Perkin, and, during the Easter holiday of 1856, he decided to investigate it in the little lab he had constructed at home. Things did not go well. When he added potassium dichromate, the chemical that was to serve as the source of oxygen, Perkin created a black goo. He tried again with aniline, a related compound, with much the same result. But then came the pivotal moment. When he mixed in alcohol to dissolve the goo, the solution turned a beautiful purple. Furthermore, the rag he used to wipe the bench was dyed the same glorious hue.

Perkin immediately recognized the importance of his accidental discovery. The color purple was all the rage in fashion circles. Manufacturers produced it either by extracting a certain species of lichen or by treating uric acid obtained from Peruvian guano (bird poop) with a series of reagents. But now Perkin could make it from cheap coal tar. He sent a sample to well-known dye makers Pullar's of Perth and requested their opinion. The answer was swift: Pullar's told Perkin that he had made "one of the most valuable discoveries that has come out for a long time." This was all the encouragement Perkin needed to leave the Royal College and — with financial backing from his father — set up a factory on a vacant lot at Greenford Green, a few miles from London.

By the time Queen Victoria appeared in a dress dyed with Perkin's mauve (as the new shade was called) at the 1862 International Exhibition in London, chemists everywhere were trying to coax novel substances out of coal tar. If Perkin could make mauve, then what other secrets did that complex mixture harbor? Soon, chemists were reporting discoveries aplenty. First came a variety of dyes; over time, these were followed by assorted drugs, plastics, and synthetic fibers, all fabricated from coal tar. William Henry Perkin's accidental discovery had set chemistry on its modern course. I would encourage anyone who is interested in delving further into the William Perkin story to read Simon Garfield's delightful book *Mauve*.

In 1906, the fiftieth jubilee of the first synthesis of mauve was widely celebrated, and among the numerous gifts that Perkin received was a painting by Arthur Cope. It's a classic. The portrait shows an elderly, bearded Perkin standing in front of some chemical glassware, brandishing a beautifully dyed piece of wool. That's the picture I use in my lectures, the one I was so eager to see in the original. Alas, it was not to be. I was informed that the portrait was not on display — it was "in storage." One of the greatest contributors to chemistry had been relegated to the basement. But a portrait of Mr. Bean was prominently displayed.

CREATOR OF GOOD AND EVIL

The green cloud rolled slowly towards the trenches at Ypres, Belgium, where the French and Algerian troops had dug in to wait for the German attack. They had never seen anything like it. And many would never see it, or anything else, again. The soldiers began to choke and cough violently as the sickly green vapor settled over the battlefield on that fateful day in 1915. Their lungs began to fill with fluid as the acrid gas stripped away their protective mucus lining. Soon, thousands lay dying, victims of a terrible new weapon. The Germans had unleashed the horror of gas warfare.

The signal to open the valves on the 5,730 canisters of chlorine gas was given by Corporal Fritz Haber, the man who had supervised the operation. That operation was deemed so successful that Haber was promoted to the rank of captain; just a year later, he would become the director of the German Chemical Warfare Service. Dr. Haber, trained as a chemist at the University of Berlin, regarded himself as a great German patriot and had accordingly volunteered his services to the military. He thought that gas warfare would be an effective way to flush the enemy from the trenches. He was convinced that it was essentially no different from shelling or bombing. The military had considered gas warfare "unsporting," but at Haber's urging they agreed to try it. On the home front, however, Haber faced stronger opposition. His wife, also a chemist, was outraged that her husband would exploit his chemical expertise in this fashion. When Haber refused to listen to her, she grabbed his revolver and killed herself. He departed the next day for the eastern front, leaving others to make her funeral arrangements. Little wonder Fritz Haber has gone down in history as the callous villain who introduced chemical weapons to the battlefield. But Haber is remembered for something else as well. He was responsible for one of the most important inventions of the twentieth century, an innovation that would save millions of people from starving to death. Talk about a paradox!

By the 1800s, scientists had begun to understand the nuances of agriculture. Carbon, which all plants need, was supplied by carbon dioxide in the air. Hydrogen came from water. But the soil supplied potassium, phosphorus, and nitrogen — minerals vital for crop growth. Once earth was depleted of these minerals, it would become infertile. One could replenish the minerals and render the soil fertile again by plowing in manure or plant wastes. Yet scientists understood that even the most efficient methods of recycling waste products as fertilizer would not be enough to sustain the world's growing population. Supplying nitrogen for crops was a particular problem. This may seem strange, since about seventy-eight percent of the air is made up of nitrogen. But, with the exception of legumes, plants cannot use nitrogen in this elemental form. Legumes have bacteria growing on their roots that can convert the gaseous nitrogen of the air into a usable form; all other crops have to rely on compounds of nitrogen in the soil.

During the nineteenth century, the major source of nitrogen fertilizer in Europe was guano imported from certain Pacific islands. Still, all the bird droppings Europeans could import were not enough to feed the growing agricultural industry. Then, as luck would have it, someone found large deposits of sodium nitrate, better known as saltpeter, in the Chilean desert. Saltpeter was a great source of nitrogen, but it wouldn't last forever. Scientists predicted mass starvation when the saltpeter ran out.

Fritz Haber turned his attention to this problem in 1904. As he saw it, the solution lay in finding a way to make use of the vast amount of nitrogen in the atmosphere. Two years earlier, Carl von Linde had succeeded in liquefying air by compressing it and then letting it expand rapidly, thereby cooling it. Through distillation, he could separate the liquid air he obtained into oxygen and nitrogen. Haber found that if he reacted this nitrogen with hydrogen under pressure, then ammonia (NH_3) would form. This was a breathtaking discovery. It meant that we could pump ammonia gas directly into soil as a fertilizer, or, even better, we could convert it into solid ammonium nitrate by reaction with nitric acid, which itself could be made from ammonia. Ammonium nitrate was an ideal water-soluble fertilizer. In 1909, Haber demonstrated his process to the chemists at the Badische Anilin und Soda Fabrik (BASF), who, under the leadership of Karl Bosch, went on to work out the final details of the industrial manufacture of ammonia. BASF signed a contract with Haber, giving him a royalty for every kilogram of ammonia produced. This made the chemist a wealthy man.

Ammonia and ammonium nitrate increased crop yields dramatically. The hungry masses could now be fed. Undoubtedly, the discovery of the ammonia synthesis was worthy of a Nobel Prize. And Haber received one in 1918 — but not without provoking controversy. Many scientists protested, insisting that the inventor of chemical warfare should not be honored with science's most prestigious prize. In 1933, when the Nazis came to power, the renowned chemist went from being a German patriot to being "the Jew Haber." Actually, Haber had never practiced his religion, and early in his career he'd even had himself baptized, joining a Protestant church to avoid anti-Semitism. Haber was not relieved of his position at the Kaiser Wilhelm Institute, but he was told he would have to replace all his Jewish scientists. He quit and departed for England, saying that he refused to judge a scientist's qualifications based on the ancestry of his grandmother.

Fritz Haber was not received warmly in England. A number of prominent Britons would have nothing to do with the father of gas warfare. And so it is that Fritz Haber leaves us with a bizarre legacy. He saved millions from hunger by synthesizing ammonia, and he killed thousands with his poison gases. German gas attacks brought gas reprisals from the Allies, which resulted in the final aspect of Haber's legacy. Gas-warfare injuries drove an Austrian corporal out of the army and into politics. His name was Adolf Hitler.

THE DARK SIDE OF RADIUM'S GLOW

I think of the Radium Girls almost every night. It happens when I check the time on my glowing watch dial — usually just after I've been reminded of the passing years by nature's nocturnal call. Let me fill you in.

In the summer of 1896, Professor Henri Becquerel went into the garden of Paris's Ecole Polytechnique and placed a photographic plate wrapped in black paper in the bright sunshine. On top of it, he carefully positioned a crystal of uranium sulfate. Becquerel was a physicist who had become interested in those amazing rays discovered just two years earlier by Wilhelm Roentgen. x-rays, as they were called, caused certain substances to glow in the dark. If x-rays could produce a glow, maybe a glowing object could produce x-rays, Becquerel thought. He knew that uranium compounds fluoresced in the sun, so he embarked on his experiment. When he developed the plate and found an exposed spot that corresponded to the spot where he'd placed the crystals, he was elated. But he still had to verify the experiment.

The next day, however, the weather was cloudy, and Becquerel put his wrapped plate and uranium in a drawer. Somehow, the plate got mixed up with some exposed plates and was developed along with them. To Becquerel's astonishment, it showed spots just like the ones uranium had produced in the sun. Sunshine was unnecessary; the uranium crystals were giving off some form of energy that exposed the photographic plate. One of Professor Becquerel's students finally determined that this novel form of radiation derived from some sort of activity within the uranium atom and was not dependent on any external stimulus. She coined the term "radioactivity." That student was Marie Curie.

Working with her husband, Pierre, Marie Curie discovered that uranium atoms were not unique in this respect. When the couple removed uranium from its common ore, pitchblende, the ore's radioactivity did not disappear. Eventually, from tons of ore, they were able to isolate a few milligrams of two new radioactive elements: polonium and radium. The latter's name was derived from the Latin word *radius*, meaning "ray," because pure radium glowed in the dark with a stunning blue color. Nobody at the time knew the reason for this eerie glow. Nobody knew that radium was spontaneously changing into radon, a process that involved the release of very energetic alpha particles. Nobody knew that it was the collision of these particles with air molecules that caused the luminescence. And nobody knew that a collision of these particles with human tissue could kill.

Radium was a novelty. Before long, entrepreneurs were busy capitalizing on its commercial potential. Glow-in-the-dark light cords flooded the stores. But what really captured the public's attention was the glow-in-the-dark watch dial. By mixing a little radium sulfate with zinc sulfide, manufacturers could produce a brilliant green glow. The energetic alpha particles released by the radium boosted the electrons in zinc sulfide into a higher energy state. When these electrons returned to their ground level, they emitted the energy they had absorbed, but this time in the form of light. We could now create luminosity without electricity.

By the 1920s, the Radium Dial Company had established itself in Ottawa, Illinois. Most of the workers were young girls, attracted by the relatively high salary of eighteen dollars a week. Indeed, the Radium Girls, as people came to call them, were easy to recognize around town because they dressed well and drove fancy cars. They also loved fun. Some took to painting various parts of their bodies with the luminous paint to surprise their boyfriends in the dark. But they were in for a surprise.

Some of the girls began to complain of jaw pain, but no one took this seriously until a twenty-five-year-old worker died. The authorities became suspicious, and they soon focused their investigations on the "lip-pointing" technique the girls had developed to paint the fine numbers on the watch dials. After virtually every stroke, a girl would lick her brush to sharpen its tip. In the process, she would swallow a little radium. Radium is insidious in that it incorporates into bone, like calcium. There, it releases alpha particles that destroy not only the bone but also blood cells in the marrow. The girls developed anemia, leukemia, and jawbones so weak that they disintegrated when a tooth was pulled. By 1929, the consequences of working carelessly with radium had become clear, but by then thirty-three workers had died. The authorities instituted reforms, ensuring that everyone wore gloves, that they mixed paint in a fume hood, and that they stored radium in lead containers. But they couldn't make the process risk-free, because they couldn't prevent radium particles from becoming airborne. The last radium-painted watches were made in 1968, just before the Radium Dial Company plant was demolished.

Now you know why I think of the Radium Girls in the middle of the night when I look at my fake Swiss Army watch, which I bought on a New York street corner for seven bucks. I doubt that it contains radium. In any case, alpha particles cannot penetrate the casing, and I do not plan to eat the watch. I suspect the zinc sulfide is energized with promethium, a synthetic element named after Prometheus, who stole fire from the heavens. It is radioactive, but it produces beta rays, which are less dangerous than alpha particles. It can be used for all sorts of glow-in-the-dark objects — maybe even toilet seats to make those nocturnal visits easier.

The Faraday Effect

Entering the grand lobby of the Royal Institution of Great Britain is practically a religious experience for anyone with a scientific bent. A portrait of Sir Humphry Davy, the institution's first leading light, peers down at the visitor over the shoulder of a statue of Michael Faraday, perhaps the greatest scientist and public lecturer who has ever lived. If you close your eyes, you can practically hear the buzzing of the crowd that filled the lobby in the 1800s. People thronged to the institution, ascended the great staircase to the lecture theater, and eagerly awaited fascinating demonstrations by Davy or Faraday.

The Royal Institution of Great Britain was the brainchild of Benjamin Thompson, perhaps better known as Count Rumford. He was by all accounts a ruthless, arrogant, cunning, devious, unprincipled womanizer who was also a philanthropist and a gifted scientist. He had invented a kitchen stove as well as a percolating coffeepot. But his most famous invention was the Royal Institution. Rumford was convinced that science should be a public servant and that people should be aware of its capabilities. The public institution he established in 1799 would be dedicated to "diffusing the knowledge and facilitating the general introduction of useful mechanical inventions and improvements, and for teaching by courses of philosophical lectures and experiments the application of sciences to the common purposes of life."

The Royal Institution became a place that not only conveyed a sense of scientific excitement to the general public through regular lectures but also served as a haven for scientific research. Funding was a constant problem, and the institution had to rely heavily on the income from public lectures. It endeavored to present lectures entertaining enough to inspire people to pay good money to attend them — in other words, the lectures had to be pretty spectacular. One of the first, delivered in 1801, featured the first-ever public demonstration of the synthesis and effects of nitrous oxide, better known as laughing gas. The presenter, Thomas Garnett, the institution's first professor of chemistry, was assisted by a young Humphry Davy, who had already made a name for himself with his work at Thomas Beddoes's "Pneumatic Institution."

Beddoes was an English physician who had noticed that cows have sweet breath, presumably due to substances in hay, such as clover. It occurred to him to take cows into the rooms of tubercular patients to purify the air with their sweet breath. Must have been quite a scene, with the patients moaning and the cows mooing. This treatment was quickly abandoned, probably because not everything the animals produced smelled sweet. Still, Beddoes was convinced that physicians could treat disease by having their patients inhale the right chemicals. He hired Davy, who had completed an apprenticeship with an apothecary-surgeon, to pursue this. Davy focused his attention on nitrous oxide. He actually recommended laughing gas as a surgical anesthetic, a recommendation that was ignored for half a century. But people were willing to experiment with nitrous oxide, and laughing gas parties became fashionable, especially after the celebrated public lecture in 1801.

Davy succeeded Garrett as professor of chemistry, and he molded the institution into a center for advanced research and public demonstrations of science. He was a peerless researcher, inventing the miner's safety lamp; that device featured a flame surrounded by wire gauze, which dissipated heat and prevented ignition of "fire damp" (methane) in mines where numerous previous explosions had occurred. He wrote the first treatise on agricultural chemistry and discovered the elements potassium and sodium when he passed an electric current through their molten compounds. A great discovery, indeed, but not Davy's greatest. Without a doubt, Sir Humphry's greatest discovery was Michael Faraday.

Faraday had no formal education. At the age of fourteen, he had been apprenticed to a London bookbinder. One day, someone brought in a volume of the *Encyclopedia Britannica* for repair, and from that day on Faraday was hooked on science. The volume included an entry on electricity, which captivated young Michael's imagination. He was absolutely thrilled when a customer offered him a ticket to one of Davy's lectures at the Royal Institution. Afterwards, he resolved to find a way to work with Davy. Faraday had taken detailed notes at the lecture, and he sent these to Davy. The great man was so impressed by Faraday's clear descriptions that in 1813 he offered him a job as an assistant at the institution. His first duty was to accompany Davy and his wife on a European tour, during which Faraday served as both scientific assistant and personal servant to Lady Davy. He didn't relish the latter, but the tour provided him with an education that he could not have found elsewhere. In the course of it, he met the leading scientists of the day, including Ampere, Gay-Lussac, and Volta, the Italian inventor of the battery.

By 1820, Faraday had become well versed in electricity, and he was ready to pounce when he heard about Hans Christian Oersted's amazing discovery at the University of Copenhagen. The Danish scientist had found that he could deflect a compass needle if he placed it near a wire through which a current flowed. This set Faraday off on a series of history-making experiments. Within a year, he had designed a device in which an electric current turned a permanent magnet — the world's first electric motor. He followed this by showing that a current in a wire could induce a current in a nearby wire — the world's first transformer. Then, in 1831, Faraday made what was perhaps his greatest discovery. He generated a current in a coil of wire as he moved it back and forth between the poles of a large magnet. The world now had the electrical generator.

Faraday had become director of the institution in 1825, at the age of thirty-four, and went on to conduct an astounding variety of experiments in addition to pursuing his work on electricity. He was the first to isolate benzene from the illuminating gas that was made by heating animal fat — the gas that Londoners used to light their homes in the early 1800s. He made new alloys of metals and produced new optical glasses. He was the first to make tetrachloroethylene, the classic dry-cleaning solvent. He liquefied many gases, allowing them to be transported in cylinders. And he gave public lectures.

In 1825, Faraday initiated the Friday Evening Discourses, which have continued to this day. One of the greatest honors that can be bestowed upon a scientist is an invitation to lecture at one of these sessions. There are some interesting traditions that go along with the honor. Before the lecture, the presenter is locked in a room with nothing but a bottle of whiskey and a chamber pot. This is because of Charles Wheatstone, a leading physicist of his day. In 1846, Wheatstone became unnerved before he was to give a lecture, panicked, and ran away. Faraday had helped prepare the lecture and was now forced to step into the breach. According to tradition, the lectures had to last exactly one hour, not a minute more or less, and Faraday ran into trouble. He finished lecturing in forty minutes; left with twenty more minutes to fill, he continued on, discussing his ideas on the relationship between light and magnetism. He had not yet refined these ideas, and he was later criticized on that score. He therefore decided that no lecturer would ever be allowed to flee again — and that explains the locked room, the whiskey, and the chamber pot.

These public lectures became so popular, especially the special Christmas lectures for children, that the street the institution stood on, Albemarle Street, became the first one-way thoroughfare in London. This eased the congestion created by the carriages bringing people to the scientific spectacles. Now, perhaps, you understand why the Royal Institution is such a special place for anyone who is interested in conveying science to the public. I had to have a picture of the famous lobby. But in order to get both the Davy portrait and the Faraday statue into the frame, I had to bow down in front of Michael Faraday. I didn't mind that one bit. It seemed the appropriate thing to do.

From Alchemist to Scientist

I learned a terrific chemical magic trick from the great English scientist Robert Boyle. He didn't teach it to me personally, since he lived roughly three hundred years ago, but Boyle was the first to record the little stunt. Onlookers were absolutely flabbergasted when, seemingly at Boyle's command, a blank piece of paper spontaneously caught fire and the flames traced out clearly legible letters. The work of the devil, they must have thought. Well, it wasn't — but it was the work of an element that some would later refer to as the devil's own. The chemical magic was performed by phosphorus.

The crafty Boyle had taken a lump of the metal from a little vial in which it sat covered with water, and he'd used it to write on the paper. As the paper dried, the phosphorus came into contact with oxygen from the air and ignited, providing a spectacular display; it was like an unseen hand writing with fire. I have always enjoyed reproducing this effect for two reasons. First, it serves as a great introduction to a discussion of the chemistry of elemental phosphorus. Second, and even more important, it marks the turning point in history from the archaic practice of alchemy to the modern pursuit of chemistry.

Although we commonly regard Robert Boyle as one of the founders of chemistry, he spent his early life searching for the philosopher's stone. This was a decidedly alchemical endeavor. The philosopher's stone was a mythical substance that could

convert metals such as iron or lead to gold. The alchemists who tried to concoct it typically recorded their experiments in secret codes understandable only to themselves. But this changed when Boyle became infatuated with phosphorus. He began to write up his observations and experiments in a way that was understandable to all. His focus shifted from trying to make gold to studying and recording the properties of matter. The secretive alchemist had undergone a transformation himself. He had become a scientific chemist.

Robert Boyle did not discover phosphorus. He learned about it from Daniel Kraft, a German who had been asked by King Charles II to come to England and demonstrate the wonderful substance that had amused the royal courts of Europe. Charles was very interested in the workings of the world, and he'd already instituted the Royal Society with the aim of promoting the study of science. Boyle was a member of the society, and he invited Kraft to his home so that he could see for himself the miraculous material he had heard so much about. Kraft wrote the word *Domini* on a piece of paper with a finger dipped in phosphorus; the letters glowed in the dark and ignited. Next he set fire to some gunpowder using nothing but a bit of phosphorus. Boyle was hooked. Kraft would not reveal the source of the substance, but after Boyle badgered and pleaded for a while, he did yield a clue. He told his host that it came from "somewhat that belonged to the body of man."

This precipitated some active research by Boyle into urine as well as into what, at the time, people referred to as "nightsoil." None of the alchemical reactions that Boyle was familiar with produced phosphorus. Ambrose Godfrey, one of Boyle's assistants, thought that he knew where they might look for help. He had apprenticed in Germany and had heard about a Hamburg alchemist who had discovered some strange substance that glowed in the dark. Boyle sent Godfrey to seek him out. From the meeting between Ambrose Godfrey and Hennig Brandt came the solution to the puzzle.

Brandt was indeed the discoverer of phosphorus. He hadn't been searching for a new element; he'd been looking for the key to immortality. Gold was considered to be an immortal metal. Unlike others, it didn't rust or tarnish. Alchemists figured that if they could isolate the property of gold that rendered it immortal, then they could use it to extend human life. Brandt's attention turned to urine. It was gold in color, and it came from the body. Maybe the body gradually lost its vitality as it lost this gold essence. In an attempt to isolate the essence, Brandt collected large amounts of urine and concentrated it by boiling it down. One day, a batch boiled dry. Brandt was shocked to see the residue glow eerily in the dark. It then burst into flame. The residue was phosphorus; in Greek, the name means "light bearer."

Brandt eventually sold his secret to Daniel Kraft, who, in turn, made a small fortune by demonstrating its properties to the rich and famous. The secret that Brandt revealed to Ambrose

Godfrey was that the urine residue had to be heated to a very high temperature to yield phosphorus. This knowledge was all Boyle needed to devise his own preparation method. Soon, he had created a supply for experimentation purposes. In retrospect, we can understand the mysteries of the urine method. Phosphates — substances in which phosphorus is chemically linked to oxygen — are plentiful in the human diet. Indeed, they are essential, because our bones are composed mostly of calcium phosphate. Molecules of DNA also incorporate phosphates, and they could not function without these units. We ingest more phosphates than the body requires, so we release the excess in our urine. Urine also contains a variety of organic substances, such as urea and creatine, which break down at high temperatures and decompose to yield elemental carbon. This carbon reacts with phosphates, stripping off the oxygen atoms to form carbon dioxide, leaving behind elemental phosphorus. Phosphorus ignites on exposure to air, a phenomenon that made Boyle's chemical magic possible.

POPPYCOCK

Not Too Hot to Handle

Have you ever had the urge to walk on red-hot coals? Maybe not. But you've probably been in a hurry to thaw a piece of meat that you forgot to take out of the freezer. Let me try to shed some light on these two activities — which, believe it or not, are scientifically connected.

You probably remember the infomercial for "Miracle Thaw" that goes something like this: "You'll never worry again about forgetting to thaw meat! Just take the meat from the freezer, put it on the miraculous thawing plate, and in minutes it's ready to be cooked. Your days of uneven microwave defrosting are over." To emphasize that the product was a true technological marvel, the infomercial showed an ice cube melting almost instantaneously on the Miracle Thaw. Should we bow our heads in reverence, or is there a more mundane explanation for this purported miracle? In order to understand what is going on here, let us digress for a moment and look at the mysterious practice of fire walking.

Who hasn't been amazed by the feat of feet treading on glowing coals? Some motivational gurus claim that one needs special powers in order to do this. And, of course, they alone are capable of teaching the mind-control techniques that protect the feet from the red-hot embers. Popular motivational speaker Tony Robbins is perhaps the most famous proponent of this idea, and he even suggests that successful fire walking is proof that his seminars have taught people to overcome extreme adversity. With the right mental focus, he asserts, we can rearrange the molecules of our feet so that they can withstand the heat. Indeed, many of his followers claim that the practice has made them feel powerful, rejuvenated, and less reliant on doctors. Robbins even says that he's seen some cancers go into remission after his fire-walking sessions. If the mind can conquer the coals, he implies, then it can also conquer disease. But you don't need metaphysical mumbo-jumbo to conquer high temperatures — plain old down-to-earth physics will do.

Ah . . . those high temperatures. Therein lies the secret. It is not only the temperature of a material that determines its potential for causing burns, but also its heat content. Temperature and heat content are very different things. Temperature is a measure of how rapidly molecules are moving about in a material. The faster they move, the higher the temperature. When rapidly moving molecules bump into slower-moving molecules, they can transfer their energy to them: the fast molecules slow down, and the target molecules speed up. What does all this mean? When we touch a hot object, the molecules in our hand speed up, and we feel heat.

But the heat we feel is dependent not only on how fast the molecules in the hot material are moving, but also on how many molecules are available to transfer their energy. In other words, it depends on the heat content of the material. A sparkler of the type used on birthday cakes serves as an illustration. The sparks that it produces are very hot, but they do not burn because they have a small mass and therefore do not contain enough molecules to cause burning by energy transfer. However, if you were to touch the sparkler itself, you'd get a nasty burn. Its temperature is the same as the sparks it sends out, but it has far more molecules ready to transfer their energy. Similarly, when you reach into an oven, your hand doesn't burn, because the mass of the air molecules in the oven is small and does not contain enough heat. If you touch the aluminum cake pan, however, you'll yowl because it has a much greater mass. Furthermore, aluminum is a very good heat conductor, so as soon as some heat transfers from the surface to your hand, it is replenished from the interior of the metal.

A knowledge of heat transfer is certainly handy — especially if you're buying jade. Experienced jade buyers can tell if a sample is real or fake just by its feel. Real jade has a high thermal conductivity, and it carries heat away from the hand very easily; fake jade doesn't. Basically, real jade is real cool.

Fire walking, on the other hand, is really hot, but it does have a cool explanation. When coals burn, their surface forms a spongy, soft layer of soot. Although this layer is quite hot, it is not massive. Relatively little energy is transferred to the foot. Also, since the coals conduct heat poorly, the heat transferred to the foot is not quickly replenished from below. Consequently, the footprints that a firewalker leaves behind are black, demonstrating that the surface of the coal has cooled down. So, it certainly isn't the rearrangement of the molecules in the feet through mind power that enables a person to walk on hot coals — it's the low heat content of spongy coal.

Now, back to our miraculous thawing device. It's really just another example of heat transfer. Think of the frozen meat as your foot, and the metal thawing plate as the bed of coals. Of course, this time we want the opposite of fire walking: we want the efficient transfer of heat to the meat. So we choose a material capable of replenishing with ease the heat transferred to the meat — in other words, a good conductor. Aluminum is a good conductor, and it's cheap. As it transfers heat to the meat, it picks up energy from the air. The thawing plate serves as a conduit of heat energy from the air to the meat. But if all this is true, then the special plate isn't really necessary. Any good conductor will do the job.

However, just to make sure that I hadn't missed some technological breakthrough, I decided to put Miracle Thaw to the test. First, I made a batch of standard-size ice cubes. Then I gathered together a variety of pots and pans. I also assembled some helpers — one wife and three daughters — for the epic experiment. Our audience was a curious cat. One by one, we placed the ice cubes on the various unheated test surfaces and, using a stopwatch, measured the time it took for each cube to melt completely. The results were conclusive. The copper pan was wonderful. The waffle iron was great. So was an aluminum frying pan. Miracle Thaw trailed the field. Even a stainless steel sink outperformed it. The overall winner? No contest. Our cat had taken an interest in the proceedings, and he decided to find out what this nonsense was all about. He began to lick one of the ice cubes, and it melted in a jiffy. Pretty good heat transfer there, but somehow I don't think we could market the cat as a meat-thawing device.

So, the secret behind Miracle Thaw and walking on hot coals lies in an understanding of temperature and heat content. If you are still in doubt, just ask a firewalker to place an aluminum sheet (or several Miracle Thaws — which have by now been relegated to the discount bin) over a bed of hot coals. The metal will reach the same temperature as the coals. Now ask the prospective walker to focus his mind, rearrange his molecules, and take the fire walk. He'll probably hotfoot it out of there real quick.

TRUE DECEPTIONS

The year is 1819. The place is the small French town of Blois. All of a sudden, the usual quiet is broken by the shrill sound of a trumpet. Everyone knows what this means: a street conjurer is about to begin his performance. Adults and children quickly gather around the performer. He mounts a bench to allow his audience a better view. Accordingly, some call him a "mountebank."

Anticipation and excitement crackle in the air as Dr. Carlosbach starts his wonderful demonstration of legerdemain. Little balls magically appear, vanish, or congregate under three inverted cups — Dr. Carlosbach is attempting a classic magic trick, called "the cups and balls." To the simpleminded onlookers, it seems as if the good doctor is achieving the impossible. And Dr. Carlosbach does nothing to discourage the belief that he has been blessed with supernatural powers. The townspeople listen enthralled as the doctor describes the other talents he possesses. He can ensure good health by destroying the vermin that infest their bodies. He can also flush out worms and distill noxious body odors. How can he do all this? Simple. The doctor will sell them some Vermifuge Balsam, a marvelous product "invented by Egyptian sages of old," for the purpose of preserving mummies.

Then, as the onlookers rush forward to grab some of the miraculous elixir, Dr. Carlosbach bursts into hearty laughter. He explains to the puzzled crowd that they've been had — the Vermifuge Balsam is nothing more than water. All he wants to do, he explains, is put them on their guard against the charlatans who routinely deceive them. For in those days the streets of French towns and villages were filled with mountebanks peddling a variety of quack nostrums to the gullible. Through his mastery of a few magic tricks, the typical mountebank was able to convince simple people that he possessed supernatural abilities.

Dr. Carlosbach felt troubled by the deception and the fraudulent use of the art and science of prestidigitation that he saw all around him. This clever street conjurer deserves recognition as one of the world's first debunkers of nonsensical therapies. But he deserves credit for something else as well. His performance on that street in Blois stirred the passions of young Jean Eugen Robert, who would go on to become one of the greatest conjurers the world has ever seen.

Magicians have been called the scientists of the stage. The wonders they perform may appear to be pure sorcery, but their tricks actually have very down-to-earth explanations — which of course they keep secret from the audience. Many of their effects involve sophisticated technology, and magicians are always keen to improve their acts by making use of the latest scientific advances. Jean Eugen Robert-Houdin (he added his wife's maiden name to his own after their marriage in 1830) is a prime example of a performer who mystified his audiences with a blend of old-fashioned prestidigitation and newfangled science.

After Dr. Carlosbach had whetted his appetite for conjuring, Robert-Houdin began to devour books on the subject. He practiced and practiced his sleight of hand, bent on becoming a magician. In the meantime, he had to earn a living, so he followed in his father's footsteps and became a clockmaker. His choice was fortuitous, because the mechanical expertise he developed would later be extremely useful in designing magic effects. In 1837, Robert-Houdin patented an alarm clock that not only woke the sleeper but also presented him with a lighted taper. He also invented the first burglar alarm, the first fire alarm (perhaps motivated by the self-lighting clock), and a variety of automata — such as "The Writer," a mechanical man that would

answer onlookers' questions in writing. By 1845, Robert-Houdin had earned enough money through these ventures to organize the first of many Soirees Fantastiques, held in Paris, which would evolve into a full-scale traveling magic show.

Robert-Houdin was the first to perform in full evening dress, unlike other conjurers of the era, who donned colorful, flowing robes and conical hats decorated with half-moons and stars. He popularized the suspension illusion — still one of the most widely performed effects in magic — by adding a new twist. Ether had just been introduced as an anesthetic, and Robert-Houdin cleverly wove it into his act. As backstage assistants poured ether onto hot shovels, flooding the theater with ether fumes, the magician explained that the chemical had the miraculous effect of making bodies defy gravity. His son then came on stage, and the magician promptly "put him to sleep" with ether. At this point, the young subject rose into the air; he remained there, suspended, with no apparent means of support. This illusion had a huge impact on people, and many of Robert-Houdin's rivals copied it. John Henry Anderson, a

Scot, performed it as "La Suspension Chloroforméen," paying homage to his countryman James Simpson, who had introduced chloroform as an anesthetic.

Robert-Houdin retired from magic after many successful European tours. He wanted to apply his energies to exploring the newly discovered phenomenon of electricity. But the government of France called him out of retirement to undertake, of all things, a political mission. Algeria was then ruled by the French, but the leaders of a religious sect known as the Marabout were attempting to stir the populace to revolt against their European oppressors. The Marabout believed that their chiefs had miraculous powers and were ready to follow them into battle. These "miraculous powers" consisted of some simple conjuring tricks that, to the uninitiated, appeared truly magical. So, in 1856, Robert-Houdin traveled to Algeria to show the people that French power was greater than Marabout power.

A French emissary invited the heads of the Algerian tribes to a special theatrical performance. That night, Robert-Houdin strode onto the stage and performed a number of his usual effects. Cannon balls magically appeared, lit candelabra were produced, and coins vanished. This aroused no great excitement — most people in the audience had seen such marvels before. Then the magician brought out a small wooden box with a brass handle, which he placed in the center of the stage. "Can anyone lift this box?" he asked. The answer was a roar of laughter. Robert-Houdin selected a particularly large and muscular audience member — a Marabout leader — and asked him to come on stage. The man easily lifted the little chest, delighting the audience. "I will now proceed to rob you of all your strength," the conjurer continued, waving his arms over the puzzled strongman. "Behold, you are now weaker than a woman!" (The World Wrestling Federation's Chyna would not be born for over a hundred years.) The grinning man bent to pick up the chest once more, but his grin soon vanished. He struggled and strained, but he could not budge the chest. Suddenly, he screamed and let go of the handle. Sweating and in a panic, he retreated to his seat.

Robert-Houdin had put his electrical knowledge to use. The bottom of the chest was made of a metal that could be attracted by a powerful electromagnet concealed below the stage. At the magician's signal, an assistant turned on the current and activated the magnet. No human could have lifted the box then. Finally, he gave another signal, and an induction coil discharged a huge shock through the handle of the chest, causing the Marabout leader to howl in pain. Then, as audience members whispered to one another about the Frenchman's power, Robert-Houdin introduced a young Moor. He asked the Moor to stand on a table and then covered him with a large cone. When he removed the cone, the Moor had disappeared. Within minutes, the frightened audience had vanished as well. The proposed revolt against France never materialized. Robert-Houdin's mixture of science and magic had saved the day for France. The great magician died of pneumonia in 1871, but his name was soon reincarnated. Erich Weiss added an "i" and became Harry Houdini. Robert-Houdin's famous namesake proved to be not only an outstanding magician and escape artist but also a scourge of fake mediums and charlatans. Just like Dr. Carlosbach.

MEDICINE AND MALARKEY

It must have been quite a scene. The little man, no more than five-foot-four, dressed completely in white, stood at center stage, playing pitch-and-catch with a chimp. But there was no ball in sight. Dr. John Harvey Kellogg was tossing pieces of steak at the chimp, who threw them right back. Then the good doctor reenacted the comic spectacle using a banana. This time, his costar didn't return the toss. To the applause of the throng that had filled the great hall at the Sanitarium in Battle Creek, Michigan, the chimp happily ate the banana. "Even a dumb animal knows what it should eat and what it shouldn't," bellowed the doctor.

Kellogg then urged those audience members who were not convinced that his performance had proved the benefits of a vegetarian diet to come on stage for a more dramatic demonstration. He invited them to gaze through a microscope at a piece of steak and a sample of manure. To their horror, they saw that the meat harbored more bacteria than the excrement! After that shocking experience, few complained about the spartan vegetable-and-grain-based diet that was standard fare at "the San."

In the late 1800s, the Battle Creek Sanitarium was the place to go if you wanted to be cured of a disease you didn't really have. Kellogg and his staff catered to rich hypochondriacs, whom they generally diagnosed as suffering from "autointoxication." The doctor was convinced that virtually all illnesses originated in the bowels, and that the "putrefaction changes which recur in the undigested residues of flesh foods" were to be blamed for disease. To cure autointoxication, you had to cleanse the bowels. In order to do this, Kellogg had at his disposal a variety of enema machines designed to flush the colon with impressive amounts of water in just a few seconds. He often boasted that he himself started the day with an enema. After flushing out a patient's nether regions with water, the doctor subjected that patient to the yogurt treatment. From both ends. Dr. Kellogg believed that the bacteria used to make yogurt were protective against disease and "should be planted where they are most needed and may render the most effective service."

There were other options for those who did not see the appeal of having yogurt pumped through their rear portals. The San's "mechanotherapy" department had come up with the "vibratory chair," a spring-loaded device that shook the patient violently to stimulate intestinal peristalsis. (See page 22.) Once toxins had been dislodged in this fashion, headaches and backaches would disappear, and, according to Kellogg, the body "would be filled with a healthy dose of oxygen." And the San's coffers would be filled with a healthy dose of money.

The San also offered a variety of baths: cold, hot, and electrified. If they didn't shock the disease out of the unfortunate victim, then Dr. Kellogg resorted to surgery, removing the offending part of the intestine. Kellogg performed over 22,000 such operations during his career, with a remarkably low complication rate. He was actually a gifted surgeon. He had trained at the Bellevue Medical College in New York, and Ellen White, the leader of the Seventh-Day Adventist movement, had financed his education. White had opened the Health Reform Institute as a center for hydrotherapy and vegetarianism, but she wanted the place to have medical legitimacy. Kellogg came from an Adventist family, and he seemed the ideal candidate to run the institute.

At the age of twenty-four, John Harvey Kellogg took up the challenge, renaming the establishment where he would practice his particular blend of medicine and malarkey for sixty-two years. The Sanitarium was a grand place. Thomas Edison, Henry Ford, S.S. Kresge, and even President Taft were visitors. They came to exercise in special athletic diapers to the beat of "The Battle Creek Sanitarium March," played by a brass band. They came to be dunked in electrified pools and to have various parts of their anatomies assaulted with streams of water. And they came for diet advice; they were told to eat what chimps eat — simple food and not too much of it. Kellogg insisted that eating meat was sexually inflammatory and would lead to "self-abuse," which robbed the body of vigor and health. He even advised his patients to curtail their sexual activity. Kellogg lived by his theories and often proclaimed that he was living proof that sex was not necessary for good health. His marriage, he said, was never consummated. We have no record of Mrs. Kellogg's views on this matter.

John Harvey Kellogg, with his brother Will Keith, developed a number of foods to replace meat in the diet. They came up with various nut butters, they were early proponents of soy, and they looked for various ways in which people could incorporate whole grains into their meals. The doctor was particularly fond of zwieback, a twice-baked biscuit that he claimed could help the bowels eliminate toxins. One day, an elderly patient broke her false teeth on some zwieback and demanded compensation from Kellogg. This prompted the brothers to cook up some wheat, pour the mush between rotating rollers, and produce wheat flakes. Cornflakes soon followed. John Harvey was only interested in the health properties of the new products. Could they serve as antidotes to the passions stirred up by meat? But Will was a businessman, and he was bent on commercialization. He eventually gained control of the family company after feuding with John Harvey, and he turned breakfast cereals into one of the world's greatest business success stories.

Dr. John Harvey Kellogg may have been eccentric, but in some ways he was ahead of his time. Menus at the San listed nutritional composition and calorie counts. He insisted that his patients get plenty of exercise and fresh air. Modern science has corroborated many of his ideas about vegetarianism, and research has shown the potential benefits of consuming foods containing certain types of live bacteria. Indeed, I've started to supplement my breakfast of cornflakes, flaxseeds, and blueberries with some live-culture yogurt. But only via the oral route.

Pi Water and Erect Electrons

Time to dip into the mailbag. I still do that, although I prefer to be questioned by e-mail. While we may worry about computer viruses, so far nobody has figured out a way to send anthrax spores via the Internet. But some people have certainly figured out how to inundate the public with scientific nonsense. One of my correspondents wanted an opinion on something called "Pi Water," which he had discovered while surfing the Web. It's a wondrous thing that slows aging, heals cuts, and — of course — cures cancer, diabetes, and hepatitis. These miracles are performed by something called "Pi energy," with which the water is infused.

As it happens, I'm quite familiar with Pi Water, since several promoters of this marvelous product have approached me. These good Samaritans wanted to ensure that the public didn't miss out on one of the greatest scientific breakthroughs of all time. Why worry about cancer and pollution when the solution is at hand? Pi Water, also referred to as "Living Water" or "High-Energy Water," is quite different from water purified by reverse osmosis or distillation, I've been told. These waters are "wet, but dead."

I must admit that I quite enjoyed learning some novel chemistry from the Pi Water "professors." I thought I knew a little bit about water, but it seems that I'm woefully undereducated. I didn't know, for example, that in the Garden of Eden, Adam and Eve drank no water. They consumed all the liquid they needed in the form of naturally grown fruits, vegetables, and herbs. Then Eve's Original Sin put an end to their carefree existence. God drove them from Eden, and they had to start drinking from ponds, lakes, and rivers — presumably due to a lack of fruits and veggies outside the garden gates. And so our downward slide from optimal health began because these waters were devoid of Pi energy, which is apparently present in the juices of plants. Things have gone from bad to worse, because today's processed water is "deader" than ever. As the Pi Water promoters explained to me, this dead water cannot carry chlorine out of our systems the way that high-energy water can. (I also learned that chlorine was originally added to water because "they didn't know what to do with the poison gas that was left over from World War I.")

But fret no more. Some wonderful technology developed by a Japanese researcher can reenergize our dead water. All we have to do is pass the water through a special filter equipped to release tiny amounts of "ferric ferrous salts" to restore its lost vigor. This energized water has an added benefit. It has been rid, I'm told, of any "memory" it may have acquired. I suppose that cruising through thousands of toilets can lead to some encounters that are best forgotten.

What are the technical details of his stupefying discovery? As I learned from one informative Web site, the astounding effects are due to "undulations of cosmic energy," presumably focused by the "ferric ferrous salts." Electrons in a water molecule start to spin faster and circle farther than they do in their usual orbit, causing them to achieve an "erected state." (I kid you not.) But the electrons cannot maintain their erected state for too long, and they try to get back to their usual orbit. When the electronic erection is lost, the cosmic energy that had been absorbed is released. These energy-creating cycles provide the wizardry behind Pi Water. And what wizardry it is! Besides curing disease, Pi Water reduces the odor of any thirsty pig that may drink it. It even makes chickens lay more, and fresher, eggs — and we certainly wouldn't want our chickens to lay old eggs. Sugar addiction succumbs to Pi Water too. A dentist writes, "the sucrose molecule combines with the dead water everyone drinks and makes it more easily drawn into our cells. Living water doesn't do this."

Is there anything positive to say about such poppycock? Well, the filters that certain entrepreneurs are selling based on the claim that they energize water may be quite suitable as simple water filters. Like any filter, they contain activated carbon, which does remove some impurities. But that has nothing to do with electronic erections. For those consumers who want the advantages of Pi Water without investing in a filter, there is a solution. In fact, it's a concentrated solution. Believe it or not, "essence" of Pi Water is available, and a few drops added to regular, dysfunctional water will viagrate it and allow it to perform like Pi Water. But the only performance I think it is capable of is raising false hopes.

HOGWASH AND BALDERDASH

Let me tell you something about cancer cells. They rush around in the blood before settling down somewhere to wreak havoc. This happens if the smallest blood vessels in the body, the capillaries, are in disorder and the cells cannot pass through them. They get stuck and begin to proliferate. To prevent the fury of cancer from being unleashed, we have to ensure that the capillaries expand so that the cancer cells can pass through. Eventually, finding no comfortable spot in which to multiply, the cancer cells are taken care of by the immune system. Capillaries expand when heated, and they contract when the body loses heat. Most of our body heat is radiated away in the form of infrared energy, having wavelengths in the eight-to-ten-micron region. So, if we generate infrared radiation in the same region and direct it at the body, then according to the principle of "complementary rays," we can prevent heat loss and reduce the risk of cancer. It's easy to do. You just have to spend thirty minutes twice daily lying in a device known as the Far Infrared Ray Hothouse, which will stimulate blood flow throughout the body. Sounds appealing, right?

Now let me tell you something else. This fascinating information doesn't come from me; it comes from the manufacturers of the Hothouse. Furthermore, it is utter nonsense. But it does a pretty good job of masquerading as science. The Hothouse looks something like a truncated doghouse that is open at both ends; it's only about a foot and a half wide. You lie inside it and experience a warm sensation, which "not only affects cancer cells but also, remarkably, helps alleviate arthritis, gout, ulcers, insomnia, aging skin, high blood pressure, pimples, and pain in the anus." It actually gives me the latter. Why? Because I find it painful to see sick people taken advantage of with false promises. I find it disturbing that the makers of a related product, a blanket, claim to use a "special light-energized bio-ceramic powder, activated by the body's own heat to produce far-infrared radiation." This balderdash just means that if you wrap yourself in a blanket you'll get warm. "Far-infrared radiation" is just technospeak for "heat." On one level, it's kind of amusing — did you know that "this energy breaks large water molecules into smaller ones, releasing trapped toxins in the process"? This may set the record for the most nonsense squeezed into the fewest words. Water molecules do not come in sizes. And they do not release toxins in response to infrared radiation.

Can this inanity be matched? You bet. By the proponents of "Energems." Like the infrared devices, these products are supposed to protect health and treat disease. Now hang on to your hats — here's the theory. The electricity that comes into your home is full of toxins that will enter your food, water, and body. This can be proven by "applied kinesiology." A kinesiologist can, for example, place a glass of water on top of a television set and then test it for toxins. To perform the test, he or she will ask a subject to stretch out one arm parallel to the ground while holding the water glass in the other hand. The kinesiologist then tries to push the extended arm down; next, the subject puts the glass down, and the kinesiologist pushes on the arm again. If the arm goes down more easily when the subject is holding the glass, it means that the poisons are at work, weakening the body. Apparently, if the TV is tuned to MTV the sample turns particularly toxic. What is the source of these electrical toxins? The U.S. Army's Ground Wave Emergency Network, a series of transmission towers spread across the country. The network creates areas of "geopathic stress." We can easily recognize such areas because in them the blood of normal individuals has a clockwise rotation; people living in stressed areas have a counterclockwise spin. This gobbledygook is certainly enough to make anyone's head spin.

Energems are the weapons we must use to counter this toxic assault on our health. They transmute electrically transmitted toxins. But what are they? To me they look like ordinary colored pebbles. Their promoters assure me that they are anything but. The Energems people have used a computer to transfer beneficial energies to the stones. You place them on the TV, the microwave, and the fridge to neutralize the electrical toxins emanating from these appliances. Be sure to tape one to your cell phone as well, because, as everyone knows, cell phone use reduces nutrient levels in our bodies — the same thing happens when we eat food that has been exposed to microwaves. Energems are also being developed for neutralizing lawn mowers and dental currents. I don't know about marketing the latter. Chewing on pebbles, computer-energized or not, doesn't sound particularly pleasant.

Energems can't solve all of our problems. If you have a lightning protection system on your house, you must remove it because it works like a cap to hold in bad energies. We know this because the dead husband of the woman who invented Energems communicated the information to his spouse through a medium. Energems do not come cheap: they cost ninety-nine dollars each. That's because they can't be mass-produced. Each one has to be appropriately energized. You cannot use your Energem pet protector to detoxify your car. Yet, mercifully, as new toxins enter the environment, you do not need to purchase new Energems. You can send your old ones in to be upgraded, for a "nominal fee."

Still, Energems are a bargain compared with the Bio-Mat system, which also improves health by reducing arthritis pain, controlling back pain, relieving stress, and expelling those ever-present toxins. Needless to say, Bio-Mat prevents cancer too. How? Well, it's made with yellow mud, natural jade, and a variety of fibers; it also emits long-wave infrared rays (those magical rays again), which penetrate the body. "There is data to suggest," its promoters say, "that cancer cells die at temperatures above 45°C." Yeah, and so do people. If you decide to fork over fifteen hundred dollars for this miracle, you will have bought yourself a very nice . . . electric blanket. I must bring this narrative to a close because I have run out of synonyms for nonsense. Okay — maybe I can come up with one more. Just think what we could do for our creaky medical system if we could funnel all the money spent on such poppycock into it instead.

HUCKSTERS SELLING HEALTH

"All sold out, Doctor!" boomed the voice that would become familiar to theater audiences across America and Europe. But in 1897, Harry Houdini was a virtually unknown entertainer with the California Concert Company, an old-fashioned traveling medicine show. As the "Great Wizard," young Harry's job was to attract and hold an audience for "Doctor" Thomas Hill, the bewhiskered pitch doctor who would stride majestically onto the stage and lecture the audience on the virtues of the miraculous elixir he had developed. As the doctor spoke, Harry and his wife, Bess, would circulate through the crowd, selling the potion. Periodically, Harry would bellow that the product was all sold out; soon, of course, more bottles would magically appear. The crowd ate up the entertainment and drank up the elixir.

The last years of the nineteenth century and the first two decades of the twentieth were the heyday of the "medicine show." This blend of entertainment and hucksterism featured an array of acts centered on the appearance of a mountebank, usually attired in a top hat and frock coat, who would pitch the product. Most would address him as "doctor," or "professor," although the only training he would typically have was from the sheep industry — the pitch doctor had to be adept at pulling the wool over people's eyes before fleecing them. This was not hard to do, because people then, as now, were eager to jump on simple solutions to complex health problems.

Many of the medicine shows featured Asians, or actors masquerading as Asians, who would nod solemnly as the pitchman delivered his lecture. These sages possessed the wisdom and healing secrets of the East, which they were willing to share with backward Americans. Even more popular than eastern miracle workers were North American Indians. People may have been suspicious of Indians, but they did believe that through living as one with nature, Native peoples had mastered botanical medicine. Indians in full regalia often led the parade as the medicine show entered town, splashing a little color into dull and drab lives.

Perhaps the most popular medicine show was the Kickapoo Medicine Show, produced by the Kickapoo Indian Medicine Company. "Doc" Healy and "Texas Charlie" Bigelow were the founders of the company, which had no connection whatsoever to the Kickapoo Indians of Oklahoma. They dreamed up Kickapoo Indian Salve as a treatment for skin diseases; the stuff was made of the "best buffalo tallow," and it was guaranteed not to contain any hog's lard. Healy and Bigelow were also pleased to introduce Kickapoo Indian Worm Killer, which they claimed would expel the parasites that caused so much human misery. And it did seem to do the job. Customers were shocked to see long stringy worms emerge from their bodies. They were stringy, all right. Worm Killer pills came equipped with their own "worms" — a length of string wound into a tight ball was packed into each pill, ready to unravel after ingestion.

However, Kickapoo Indian Sagwa was destined to become the main nostrum of the hundred or so Kickapoo Medicine Show companies that toured America. Healy and Bigelow insisted that Sagwa was a cure-all, and they even got Buffalo Bill Cody to endorse it in ads: "An Indian would as soon be without his horse, gun, or blanket as without Sagwa." Indians, of course, were always without Sagwa. Healy and Bigelow had invented the product and the name. Some of the advertising was more insidious. Then, as now, hucksters peddled their products by casting aspersions on medications prescribed by physicians. "Poisoned by Calomel — Cured by Sagwa" proclaimed an ad that took direct aim at a widely prescribed cathartic medication. Calomel (mercurous chloride) was not a great medication, to be sure, but no problem it caused could be cured with Kickapoo Sagwa.

Most concoctions hawked at medicine shows were harmless — and useless — brews made from herbs, roots, and bark. They usually contained a hefty dose of alcohol, which increased the potential for customer satisfaction. Some even contained a little opium to brighten the mood. Medicine makers often mixed a laxative into their potions as a means of demonstrating to their customers that toxins were truly being expelled from their bodies.

Medicine shows prospered because most people did not realize that many ailments are self-limiting; nor were they aware of the pitfalls of anecdotal evidence or the power of the placebo. But in 1906, the U.S. government passed the Pure Food and Drug Act, which required manufacturers to list ingredients on patent medicines and to curtail the hype surrounding these products. The pitches were toned down, the products were demystified, and the medicine shows rapidly lost their appeal. By 1914, even the Kickapoo road companies had stopped battling their fate; they buried the hatchet and disbanded. But the medicine shows had not vanished forever. And when they reemerged, it would be science that apparently legitimized them.

During the first half of the twentieth century, vitamins replaced Indian remedies as magical cures in the eyes of the public. Scientists had shown that devastating diseases such as rickets, pellagra, scurvy, and beriberi responded to vitamin therapy, a finding that provided fodder for hucksters, who launched a barrage of wild claims. Senator Dudley J. LeBlanc of Louisiana established Happy Day, a company that would become famous for its star product: Hadacol. The "happy day" was the day a doctor had injected LeBlanc with B vitamins to treat some pain he had been having. The senator was so thrilled with the results that he developed Hadacol — a mixture of B vitamins and iron — to share with the world. LeBlanc was a master salesman. He published testimonials in newspapers; in them, people claimed that Hadacol had alleviated their arthritis, asthma, diabetes, epilepsy, heart trouble, tuberculosis, or ulcers. When the Food and Drug Administration got after him, LeBlanc toned down the claims and cleverly suggested that Hadacol was good for what ailed you, as long as what ailed you was what Hadacol was good for.

In 1950, LeBlanc reinvented the traveling medicine show. A caravan of 130 vehicles toured the South, entertaining as many as ten thousand people a night, all of whom had come with Hadacol box tops as admission tickets. They were treated to musical numbers like "The Hadacol Boogie" and "Who Put the Pep in Grandma?" Mickey Rooney, Chico Marx, and Burns and Allen performed comedy skits; clowns took long drinks from Hadacol bottles, and their false eyes and noses lit up. Hucksters offered amusing quips about Hadacol's powers — like "Have you heard about the ninety-five year old who was dying in the hospital? She was taking Hadacol, but it didn't save her. Did save the baby, though."

The fame of Hadacol spread, and so did the profits. LeBlanc was spending an amazing one million dollars a month on advertising and grossing twenty million a year. The man even had a sense of humor. On a talk show, Groucho Marx asked LeBlanc what Hadacol was good for. LeBlanc retorted that "it was good for five and a half million for me last year." But within a few years, the public had figured out who was benefiting the most from Hadacol, and soon the product was history.

Today, the traveling medicine show, with its fascinating mix of fun and flimflam, is gone. But not forgotten. If you get a chance, take in a psychic fair or a health food expo at a hotel or convention hall, and experience a throwback to the past. I did. There were crystal healers, astrologers, and dietary supplements galore. Some of the claims sounded like they came straight from the mouths of the Kickapoo pitch doctors. One booth even had a magician performing tricks to attract a crowd to a display of a wondrous new nostrum. He was not very successful. His magic was roughly on a par with that claimed for the product. He was certainly no Houdini.

INDEX

About the Author

Joe Schwarcz is a professor of Chemistry and the Director of the Office for Science and Society at McGill University in Montreal, Canada. He hosts a popular weekly phone-in radio show, is a regular on Canada's Discovery Channel, frequently gives entertaining and educational public lectures, and writes a column for the Montreal *Gazette*. He has received many honors, including the American Chemical Society's prestigious Grady-Stack Award for Interpreting Chemistry for the Public. He lives in Montreal with his wife and three daughters.